D1739668

Global Development and the Environment Series

Series Editors
Richard M. Auty and Robert B. Potter

Land Degradation in the Tropics

Titles previously published in the Global Development and the Environment series, under the Mansell imprint:

Economic Development and the Environment
Agricultural Change, Environment and Economy
Economic Development and Industrial Policy
Water and the Quest for Sustainable Development in the
Ganges Valley

Land Degradation in the Tropics
Environmental and Policy Issues

Edited by

Michael J. Eden and John T. Parry

PINTER

First published in 1996 by
Pinter, *A Cassell Imprint*
Wellington House, 125 Strand, London WC2R 0BB, England
127 West 24th Street, New York, NY 10011, USA

© Michael J. Eden, John T. Parry and the contributors, 1996

All rights reserved. No part of this publication may be reproduced or transmitted in any form
or by any means, electronic or mechanical, including photocopying, recording or any
information storage or retrieval system, without permission in writing from the publishers or
their appointed agents.

British Library Cataloguing-in-Publication Data
Land Degradation in the Tropics : Environmental and Policy
    Issues. – (Global Development and the Environment)
    1.Land Degradation – Tropics 2.Environmental Degradation –
    Tropics
    I.Eden, Michael, 1936 – II. Parry, John T.
    333.7'3137'0913

ISBN 1–85567–389–4

Typeset by York House Typographic Ltd, London
Printed and bound in Great Britain by
Biddles Limited, Guildford and King's Lynn

Published with the assistance of The Commonwealth Foundation

THE
COMMONWEALTH
FOUNDATION

Contents

Part III Degradation in the Drier Tropics

Part IV Degradation in Tropical Wetlands

Part V Urban and Industrial Degradation in the Tropics

Part VI Conclusion

List of Tables

List of Illustrations

Contributors

Chebo K.A. Asangwe,
Department of Geography and Planning, University of Lagos, Nigeria

Christopher J. Barrow,
Centre for Development Studies, University College of Swansea, Wales

Kawi Bidin,
School of Geography, University of Manchester

Tanya A.S. Bowyer-Bower,
Department of Geography, School of Oriental and African Studies, University of London

Naresh Chandra Gautam,
National Remote Sensing Agency, Balanagar, Hyderabad, India

Michael B.K. Darkoh,
Department of Environmental Science, University of Botswana, Gaborone, Botswana

Ian Douglas
School of Geography, University of Manchester

Michael J. Eden,
Department of Geography, Royal Holloway, University of London

Jennifer Elliott,
Department of Construction, Geography and Surveying, University of Brighton

Alan Grainger,
School of Geography, University of Leeds

Tony Greer,
Department of Geography, National University of Singapore, Singapore

Sally Lloyd Evans,
Department of Geography, University of Reading

Frank McShane,
Department of Geography, McGill University, Montréal, Canada

Ravoori Nagaraja,
National Remote Sensing Agency, Balanagar, Hyderabad, India

John T. Parry,
Department of Geography, McGill University, Montréal, Canada

Robert B. Potter,
Department of Geography, Royal Holloway, University of London

Waidi Sinun,
Innoprise Corporation Sdn. Bhd., Kota Kinabalu, Sabah, Malaysia

John A. Soulsby,
Department of Geography, University of St Andrews, Scotland

Jadda Suhaimi,
School of Geography, University of Manchester

Azman bin Sulaiman,
School of Geography, University of Manchester

Patrick Williams,
Department of Geography, University of Guyana, Georgetown, Guyana

Lovemore M. Zinyama,
Department of Geography, University of Zimbabwe, Harare, Zimbabwe

Foreword

The principal aim of the Global Development and the Environment series is to provide an outlet for scholarly work covering important aspects of Third World development and change. The series is aimed at a multi-disciplinary audience and it is intended that the issues covered will be treated from a variety of different disciplinary perspectives – economic, social and political, historical and environmental among them. At the same time, we are aware of the need to achieve balance with respect to the various regions of the Third World that are covered by the volumes in the series, not least because of the striking heterogeneity that is so characteristic of the nations that make up what we refer to in shorthand terms as the 'developing world'. In essence we are seeking to promote the publication of works that deal in a rigorous manner with Third World themes and issues that are of topical interest and pressing social importance.

One important objective of the series is to encourage new and bold perspectives on development problems and issues. A second key objective is to develop inter-country comparisons that achieve balanced coverage of the principal regions of the world. It is hoped that such inter-country comparisons will shed new light on the ways in which differing social, cultural, economic, political, ecological and natural resource systems condition responses to global processes of change.

The series is thereby built around two closely related themes: globalization and environmental change. Globalization is a major trend affecting contemporary Third World countries. It is reflected in the diffusion of capital and technology, the evolution of new production systems and the spread of Western life-styles among elites and other groups. It is also witnessed in the increasing importance of multinational corporations. Yet it is clear that the processes of global restructuring and change are affecting various regions and nations at different rates and in a variety of different ways. For example, large income gaps have opened up within countries of Latin America and the Caribbean, while pressures on resources vary markedly among the various rural areas of Sub-Saharan Africa. Similarly, rates of economic growth have diverged sharply in East Asia. It is clear that patterns of production are becoming increasingly heterogeneous when viewed at the international level, while patterns of consumption and associated aspirations are frequently converging on what might be described as a global norm. However, such patterns of consumption are likely to be found to be strongly differentiated when they are examined in different groups and areas at the local scale.

Environmental change is strongly affected by the globalization process, whether through the clearance and destruction of rain forest, the occurrence of industrial accidents, the despoliation of attractive environments, indigenous cultures and socio-economic landscapes by the demands of international tourism, or the consequences of global warming for sustainable patterns of development and resource use. The examination of the interacting socio-political and environmental causes of these problems, and the practical responses, stands as a further major theme of the series.

The present volume, jointly edited by Michael Eden and Professor John Parry, tackles the urgent problem of land degradation in the tropics. It touches on a spectrum of issues which are strongly in line with the main objectives of the Global Development and the Environment series. The environments which are analysed range from rain forests through tropical dry forests and wetlands to cities, and examples are drawn from all the principal cultural regions which make up the developing countries. The book also welds together analyses of the technical problems posed by the need to understand and measure environmental change along with evaluations of policy responses from a variety of different perspectives. It is only out of such an interdisciplinary matrix that the complex problems of scientific measurement, policy formulation and effective policy implementation which sustainable development requires will be resolved. This book is, therefore, a very welcome addition to the series.

Rick Auty   Rob Potter
Lancaster, Lancashire   Englefield Green, Surrey
England   England

Preface

The problems of land degradation are increasingly apparent. During the long tenure of the earth by humans, terrestrial environments have suffered localized damage and degradation. In the present century, these impacts are more sustained and far-reaching. Early warning signals went largely unheeded. In the mid-nineteenth century, George Perkins Marsh wrote eloquently of the damage being done to the woods, waters and soils of his native Vermont. Marsh's book *Man and Nature*, published in 1864, is rightly seen as the foundation document for the conservationist movement in America. It was to have an important effect on government policy in relation to habitat and wildlife management. Unfortunately, the twentieth-century revolution in engineering and technology ran roughshod over the warnings of the early conservationists. Technocrats proclaimed a future world in which nature could be manipulated and plundered. It has taken the greater part of the twentieth century to realize that massive engineering intervention and a robber economy all too often lead to devastation.

While the Western world is slowly coming to terms with the need for a new *rapprochement* between technology, development and the environment, it is clear that in the tropics the driving force for rapid development and industrialization is powered by policies of exploitation for quick gain, and, in all too many instances, the culprits are the capitalists of the Western world. Tropical countries are now having to deal with the legacies of exploitation and environmental mismanagement, and at the same time cope with the processes of global change, such as climatic warming, sea-level rise and desertification.

The Commonwealth Geographical Bureau and its Committee of Management have been concerned with these issues for some time. In 1993, an invitation was issued to Commonwealth geographers to contribute both case studies and broader perspectives on the problems of land degradation in tropical forests, tropical drylands and wetlands, and the rapidly expanding conurbations of the tropical world. The result is the present book, its twenty-two chapters ranging across the spectrum of land degradation with a focus on policy issues and on management strategies. It is hoped that the book will be a worthy companion to earlier publications of the Commonwealth Geographical Bureau devoted to tropical themes. Funding in support of publication has been received from the Commonwealth Foundation and the editors wish to record their appreciation of this grant.

The book is organized in six parts. In the introductory chapter, the broad issues of land degradation are outlined and important social and policy issues for developing countries are addressed. Radar remote sensing is discussed in the second chapter, and it is shown that spacecraft radar systems can provide high-resolution spatial data for a wide range of land degradations even in areas of persistent cloud. In addition, the re-visit capability of orbiting satellites offers a means of updating and monitoring rapidly changing situations.

In Part II, four chapters are devoted to the serious problems of tropical forests. Following a general review of the types of degradation affecting these forests and the issues of damage control faced by most tropical countries, three regional studies provide insights into specific sets of problems. In the first of these, the general human impact on the Amazonian forest is examined with particular reference to forest renewability. The forest lands of Southeast Asia have received relatively little attention from conservationists, but many of the problems and issues are particularly serious in this area, as explained in the second regional study. In the third study, logging practices in Sabah are assessed as they relate to soil loss in a high rainfall area.

Part III of the book is devoted to the drier tropics. In a general introduction, fundamental problems, such as population pressure and overgrazing, are set in a broad framework. Of the five other chapters in the section, four are concerned with land degradation in tropical Africa. Examination of a practical method for assessing soil erosion using Landsat imagery for Malawi is followed by an assessment of the pressures on land in resettlement areas in Zimbabwe. Some of the conflicting issues of land management in Kenya are examined from two perspectives – that of regional and local interests and that of central government. Similar problems of environmental management are identified in the last of the African contributions, which deals with contradictions in government policies in Zimbabwe. In the other contribution in this section, two researchers at the Indian National Remote Sensing Agency describe the application of remote sensing to identifying, mapping and monitoring wasteland in India.

In Part IV, some of the problems of tropical wetlands are addressed. At present, wetland development proceeds at a rapid pace, particularly when the wetlands are adjacent to urban centres. In the introductory chapter, the special characteristics of wetland ecosystems are examined and some examples of wetland conversion and reclamation assessed in relation to short-term and long-term cost benefits. In the following chapter, the particular problems of riverine habitats in Amazonia are examined. In many parts of the tropics, coastal mangrove forests have been removed for aquaculture or rice schemes. The effects of destruction and degradation of mangrove are illustrated in a comparative study of Belize and Fiji. In the final chapter, the impact of rapid urbanization on a coastal wetland is examined in a case study of Lagos, Nigeria.

Urban and industrial impacts may well be the most serious of all types of land degradation in the tropics and these are examined in Part V. In the introduction, an overview of problems faced by the fast-growing cities of the Third World shows the seriousness of the situation. In the three chapters that follow, the specific problems of two capital cities, Harare and Georgetown, are examined, and there is a broader review of the environmental impacts of urban development in the Caribbean. The final chapter in the section focuses on a proposed mining project in Fiji, which, because of its size, the land rights issues and the societal problems of a high-wage mining development, is likely to generate serious conflicts for a small island community.

In the final section, the editors attempt a brief synopsis of land degradation in the tropics and identify some of the strategies for minimizing environmental damage. Technical and political issues are examined and the value of a broadly based, geographical approach to the problem is demonstrated.

J.T. Parry
Montréal

M.J. Eden
London

PART I

Introduction

1

Land Degradation: Environmental, Social and Policy Issues

Michael J. Eden

It is easy enough to be pessimistic about the environmental prospects of the tropical world. Especially since the emergence of the global environmental movement in the late 1960s, repeated concern has been expressed over degrading land conditions in the tropics and their impact on human welfare. In this respect, non-governmental organizations and the media have often taken a lead from pioneer researchers, who have themselves frequently campaigned against tropical environmental mismanagement and flawed development schemes.

In earlier decades, such campaigning was especially productive in that it drew attention to tropical environmental issues and alerted individuals, agencies and governments to the hazards of new, intensifying or extending resource uses and land developments. In particular, it contributed to the growing political awareness of the global environment that was symbolized by the United Nations Conference on Environment and Development in Rio de Janeiro in 1992. In respect of tropical forests, for example, seminal contributions were those of Gómez-Pompa *et al.* (1972), who promoted the idea of the 'non-renewability' of the forest, and of Richards (1973), who predicted its destruction, save for a few small reserves, by the end of the century. Neither view is acceptable today, but both have been widely cited over the years and been influential in alerting a global audience and provoking further research.

Previous concern with tropical environmental issues has mostly been system-specific and reductionist in approach. Alongside deforestation, which has been in the headlines for two decades, other issues have received attention, including laterization and soil erosion, soil nutrient depletion, waterlogging and salinization, savannization, desertification, loss of bio-diversity, urban–industrial pollution and the like. While some of these processes are not exclusively tropical and some are less serious than others, many of them continue to be treated as discrete environmental issues in the tropics. Admittedly, desertification has developed into a broader concept encompassing linked climatic, edaphic and biotic processes, and is nowadays seen as the major environmental problem of arid or semi-arid areas such as the African Sahel (Verstraete, 1986; Adams, 1990). Latterly, there has also emerged a more general concept, land degradation, which is the subject of this book and which, in many respects, offers an enhanced perspective on the environmental issues mentioned above.

The present chapter attempts a broad review of land degradation. Firstly, the emergence and definition of the concept are outlined. Secondly, its nature is examined in terms of underlying social factors and constituent bio-physical processes. Thirdly, some policy aspects of land degradation are explored and are discussed in relation to its management.

Concept of Land Degradation

It has long been recognized that degradation of the bio-physical environment is a process induced or accelerated by humans (Schreckenberg *et al.*, 1990), but the term is one of a number that have been used in this context. Thus, periodic references occur to 'degeneration of the soil' (Nye and Greenland, 1960), 'exhaustion of the land' (Gourou, 1961) and 'environmental deterioration' (Dasmann *et al.*, 1973). The term degradation has also long been used, but in a narrower rather than broader context. References exist to 'degradation of the structure of the soil' (Nye and Greenland, 1960), 'degraded soils' (Dasmann *et al.*, 1973) and 'degradation of tropical vegetation' (Manshard, 1974). However, land degradation has latterly emerged as a more formal, generalizing concept that covers not only the dynamics of soil, vegetation and other perturbed natural systems, but also related social impacts.

An early expression of the broader approach was that of Chisholm and Dumsday (1987), whose edited volume *Land Degradation: Problems and Policies* was specifically concerned with Australia. It focused on issues of regional degradation ranging from soil erosion and salinization to recreational damage and loss of habitats. Its main emphasis was on socio-political rather than bio-physical elements of the subject. A contemporary but more ambitious treatment was Blaikie and Brookfield's (1987) edited volume *Land Degradation and Society*, which stressed the value of an integrated environmental and socio-political approach. Land degradation was seen as an interdisciplinary issue and one that 'should by definition be a social problem' (Blaikie and Brookfield, 1987). The approach was global in character and, although the systematic perspective was somewhat narrow, with recurrent emphasis on soil erosion, the strength of the volume was its attempt to create an integrative theoretical framework for land degradation and explore implications for land management.

Another contribution was Barrow's (1991) volume *Land Degradation. Development and Breakdown of Terrestrial Environments*. This was also global in perspective. Its approach was empirical rather than theoretical, with an emphasis on technical issues, but it again stressed the value of integrating environmental and human elements of land degradation. Likewise, in Johnson and Lewis's (1995, p. 23) volume *Land Degradation: Creation and Destruction*, land degradation was seen as 'a response to a multitude of complex interacting physical processes along with human values and constraints'. In general, any attempt to integrate these variables is extremely complicated, but it is at least now widely recognized as a desirable objective. The approach is also evident in the parallel concept of 'critical environmental zones', which, in 1992, was formally established as a theme of a study group (now a Commission) of the International Geographical Union. Among the study group's concerns were the notions of environmental 'endangerment' and 'criticality'. The aim was that of 'treating the driving forces responsible for human-induced environmental criticality and assessing how societies respond to such changes' (Kasperson and Heffernan, 1992, p. 2).

The drive for an integrated approach to land degradation contrasts with the traditions of many environmental and social scientists, whose perspectives have often been excessively blinkered. Nugent's (1990) assertion, for example, that the Amazonian forest is 'part of a complicated social landscape' as well as being an ecosystem aptly reminds us of the shortcomings of much previous ecological research in that region. In a broader Third World context, Redclift (1987) acknowledges a converse neglect of the environment by most social scientists. Bridging the gap is facilitated by the concept of land degradation, even though it creates theoretical and methodological problems, not least in terms of matching environmental or technical solutions to social or economic realities. As Bojo (1991, p. 75) indicates, 'physical data on land degradation are of little use to decision-makers unless transformed into units comparable with the cost of soil conservation'.

The broader approach to land degradation is symbolized by Blaikie and Brookfield's (1987) assertion that land degradation should be seen as a 'social problem', even though formal definitions generally focus on the land itself: 'when land is degraded, it suffers a loss of intrinsic qualities or a decline in capability' (Blaikie and Brookfield, 1987, p. 6). Johnson and Lewis (1995, p. 2), although again identifying 'human interference' as the cause of land degradation, likewise define the latter as 'the substantial decrease in either or both of an area's biological productivity or usefulness'. A further important point is that of Blaikie and Brookfield (1987) who identify the process of *net degradation*, which recognizes not only the coexistence of human interference and natural degrading processes, but also parallel 'natural reproduction' and 'restorative management'. Any relevant land-management policies must reflect this approach.

The concept of land degradation has emerged from what is widely perceived as an overpopulated or overexploited world. As such, it is a belated response to the global environmental movement that arose in the late 1960s and extended specifically to the tropics in the 1970s. Land degradation itself, however, is no new phenomenon in the tropics or elsewhere. Its scale and intensity have greatly increased in recent decades, but, on a timescale of centuries, it is possible to identify a similar acceleration of destructive land exploitation that coincided with European colonialism (Manshard, 1974; Blaikie and Brookfield, 1987). Nor should exaggerated devotion to the idea of tribal adaptation to the environment obscure recurrent evidence of at least localized, pre-modern land degradation on a timescale of millennia. Tropical island environments, notably in the Pacific and the Caribbean, suffered early biotic and edaphic degradation (Steadman *et al.*, 1984; Nunn, 1990; Steadman and Kirch, 1990; Diamond, 1991; Bahn and Flenley, 1992). Likewise, pre-modern forest clearance and soil erosion occurred in mainland areas of the Old and New World tropics (Prance and Schubart, 1978; Hamilton *et al.*, 1986; O'Hara *et al.*, 1993).

It is evident that in spite of increasing contemporary land degradation, paralleled by growing population densities and advancing technologies, land degradation is a complex phenomenon occurring under diverse social and environmental conditions. Its solution is also complex, lying as much with policy formulation and land management as with field experiment and laboratory analysis.

The Nature of Land Degradation

Investigations of specific kinds of land degradation have long been undertaken in the tropics. They have involved both rural and urban–industrial environments and have focused on perturbed physical, chemical and biotic systems. Although discrete investigations of specific degradations, like soil nutrient depletion (e.g. Falesi, 1976; Serrão *et al.*, 1979) or savannization (e.g. Reiner and Robbins, 1964; Eden, 1974), have frequently occurred, linkages between bio-physical processes are apparent and some degradations overlap. This is particularly the case with desertification, which has been notable for its broad application. In the past, the term has been used in the humid tropics where deforestation has been seen as a potential precursor of desert-like conditions (Aubréville, 1949; Goodland and Irwin, 1975), although its current usage is in arid to semi-arid areas like the Sahel. Even there, it embraces the phenomena of drought, soil erosion and loss of vegetation cover as well as the degradation of human living conditions (Verstraete, 1986).

Comprehensible reasons exist for discrete technical investigations of specific degradations, but, as recent desertification studies show, other benefits accrue from broader and more integrated treatments (Adams, 1990; Grainger, 1990). This equally applies at the level of land degradation itself, where, in spite of the multiple bio-physical processes involved, common underlying social causalities often exist whose comprehension is necessary for relevant policy intervention. Likewise, the bio-physical impacts of degradation not only create technical problems that need solution, but also affect the economic activity and social welfare of human populations. As Young and Ishwaran (1989, p. 10) indicate, 'it is no accident that poverty and degraded environments are generally coincident'.

In practice, this means that land-management policies must be appropriate in human as well as technical terms. Self-evident as this may be, it is often neglected in technically orientated studies of specific degradations. In purely bio-physical terms, benefits also accrue from an integrated approach to land degradation in that it encourages scrutiny of the linkages between specific degradations, reveals commonalities in degradation monitoring, and facilitates assessment of the relative criticality, and hence political significance, of particular degradations.

Social Aspects of Land Degradation

Consideration of the view that different land degradations have similar causes, relating for example to the chosen modes, intensities or technologies of land use, rapidly leads to consideration of more fundamental and controversial problems of human behaviour and interaction with the environment. As indicated, land degradation in the tropics is an ancient phenomenon and, although its impact has increased over time along with human numbers (Myers, 1992), population density is by no means its exclusive cause. Admittedly, a significant demographic influence is canvassed and the potential environmental benefits of slower population growth are often emphasized, as in the Brundtland report (World Commission on Environment and Development, 1987) and at the International Conference on Population and Development in Cairo in 1994 (Dickson, 1993; Sadik, 1994). But land degradation is not uniquely associated with densely populated land, and is nowadays as likely to derive from commercial land use that is anything but labour-intensive like cattle ranching in Amazonia or extractive logging in Southeast Asia.

In such instances, it is the mechanization of land exploitation, symbolized in the rain forest by an historic progression from the stone axe, through the machete and the steel axe, to the chainsaw and bulldozer, that is more directly responsible for land degradation than sheer numbers of people. As Ohlin (1992, p. 8) claims, 'it is not difficult to show that the bulk of environmental stress is due to technology rather than population growth'. The case is arguable, although it must be recognized that ultimately technology is neutral and no more than the tool of the land user who employs it. Even so, a failure to come to terms with new technology, mechanical or chemical, commonly increases environmental degradation.

Equally critical, and very relevant in the tropics, is the issue of access to land and of property rights, which in turn affect *de facto* population densities, local land uses and hence environmental impacts. In the tropical Andes, for example, overcrowded smallholdings are widespread and commonly degraded; they are often unproductive and cultivated impermanently. Their existence reflects inequitable land distributions and the failure to achieve proper land reform in a region where more fertile tracts have long been held in large, under-used units (Watters, 1971; Gilbert, 1974). Structural factors of this kind also contribute to land degradation.

Other structural factors are influential in so far as commercial entities, including multinational companies, are widely involved in tropical land development and add to land degradation. Logging and ranching operations have been mentioned, while large-scale mining activities also induce biophysical degradations that affect local populations. Economic benefits may accrue, but the attendant social costs are often high. Notable examples include the copper mining on Bougainville island and at Ok Tedi in Papua New Guinea which have damaged local communities and their land (Brown, 1974; Hurst, 1990; Bordia, 1993).

Broader development strategies pursued by international agencies or by developed world governments via aid programmes have also had serious environmental and social impacts that have been ignored or neglected in anticipation of economic benefits (Hayter and Watson, 1985; Rich, 1994). Recurrent examples exist in Brazilian Amazonia, not least in respect of transport and hydroelectric projects associated with the *Programa Grande Carajás* in Pará or the *Programa Integrado de Desenvolvimento do Noroeste do Brasil* in Rondonia (Hall, 1991; Shankland, 1993). In the face of pressure from non-governmental organizations during the 1980s, institutions like the World Bank have sought to improve their environmental performance (World Bank Group, 1994), but enhanced policies based on sustainable development and low environmental impact are not easily implemented and criticism of such institutions has persisted, not least in respect of structural adjustment policies (Prowse, 1994; Rich, 1994).

Land degradation in the tropics thus has diverse underlying causes that are broadly social in character and often far removed from the actual land user. Land degradation is as likely to arise from short-sighted policies or an incompetent performance on the part of international agencies or national governments as from ignorance, neglect or misplaced priorities on the part of the actual land user. From this pattern of causality derive human impacts that in turn damage the land user and the broader society. In effect, parallel social degradation occurs. Irrespective of whether the initial land degradation involves, say, soil erosion or salinization or arises in the humid or semi-arid tropics, recurrent patterns of human impoverishment occur that are expressed in terms of food shortage, ill-health, declining income, inadequate housing and the like. Such impoverishment is widespread in rural areas, but also exists in towns and cities where the proximate causes range from industrial pollution to inadequate waste-disposal systems. Urban environmental degradations in the tropics have received less attention than their rural equivalents, but, in terms of human numbers, the potential impact is greater and increasing rapidly as the urban–industrial environment itself expands.

Technical Aspects of Land Degradation

An integrated approach to land degradation yields technical benefits in respect of environmental planning and management. The approach emphasizes the systemic and spatial linkages between individual degradations, and, in particular, encourages a sharper investigation of the connections between terrestrial, aquatic and atmospheric systems and of the broad degradations relating thereto. The latter include the assumed climatic and hydrologic feedbacks of large-area land cover changes and associated losses of biodiversity. Such bio-physical processes have global as well as tropical implications, but remain poorly documented and inadequately understood.

It is assumed, for example, that tropical deforestation, as well as causing local soil and vegetation degradation, lowers evapotranspiration rates and hence regional rainfall levels. Simulation modelling confirms this (Lean and Warrilow, 1989; Shukla *et al.*, 1990), but field data are limited and doubts exist regarding critical model variables, not least the assumed post-clearance land cover (Eden, Chapter 4). Even broader linkages have been postulated between deforestation in the humid tropical zone and desertification processes in the Sahel, but again little direct evidence is available (Hulme, 1989). Large-area deforestation is also assumed to have damaging downstream effects on fluvial systems, particularly in respect of flood regimes, turbidity and biodiversity. Again the complex linkages are poorly understood, and, in respect of both the Himalayan region (Ives, 1991; Hasan and Mulamoottil, 1994) and Amazonia (Richey *et al.*, 1989), in need of more careful appraisal.

An integrated approach to land degradation can also promote more effective environmental monitoring. Local studies have been undertaken (e.g. Kummer, 1992), but the use of satellite imagery to monitor large-area degradations like deforestation in Amazonia or desertification in the Sahel has been disappointingly slow and only lately begun to realize its potential (Helldén, 1991; Tucker *et al.*, 1991; INPE, 1992; Martini, 1992). Likewise, broader attempts at baseline surveys of land degradation itself are only now beginning to appear, as in Mexico (Anon., 1992) and India (Nagaraja and Gautam, Chapter 11).

Various reasons exist for the delayed application of satellite imagery to large-area land degradation monitoring (Parry, 1986), but the value of a regional or pan-tropical overview of degradation-related phenomena is increasingly recognized, and monitoring systems with more appropriate spatial, temporal and spectral resolutions are becoming available. Neither Landsat nor NOAA AVHRR imagery has been ideal for tropical environmental monitoring, but the appearance in the 1990s of cloud-penetrating, satellite radar systems like ERS-1, JERS-1 and RADARSAT (Wooding and Attema, 1994) and plans for a Moderate Resolution Imaging Spectrometer (MODIS-N) (Townshend *et al.*, 1991) will contribute to monitoring tropical land cover and degradation. As well as providing information for governments in the tropics, the satellite data will be of use to international funding agencies, non-governmental organizations and other bodies.

An integrated approach to land degradation can also encourage more objective assessment of the *relative* impact, and hence socio-political importance, of specific degradations. Hitherto, the tendency has been to identify fragile environments and potential land degradations in all directions, usually on the basis of postulated 'worst-case scenarios', when in reality some degradations are more serious than others in terms of the inherent fragility of the natural systems concerned, the scale or intensity of the impacts involved, and the reversibility of those impacts. In recent years, much attention has focused on desertification and deforestation, while environmental degradation associated with urban–industrial areas has received much less attention in spite of the large and increasing numbers of people involved. Inadequate investigation of the relative seriousness or criticality of individual degradations creates a major problem for governments in the tropics who often have limited resources and an uncertain commitment to the environment. The need exists for a balanced view of the relative criticalities, so that rational environmental policies appropriate to the available investment can be formulated by national governments or external funding agencies.

As is evident from Blaikie and Brookfield's (1987) emphasis on net degradation, such policy-making needs to include consideration of the relative reversibility of degradation processes and the relative renewability of degraded landscapes. Much attention has been paid, for example, to nutrient loss in tropical soils because of the prompt impact this has on crop yields, while soil physical changes, especially soil erosion, have been relatively neglected in spite of the greater difficulty, on a human timescale, of renewing eroded as opposed to leached soils (Ahn, 1970; Richards, 1985). Similarly, abundant concern has been expressed over deforestation without adequate appraisal of forest renewability and its implications for anticipated regional bio-physical feedbacks (Brown and Brown, 1992; Skole and Tucker, 1993; Eden, Chapter 4). More careful assessment is thus required of degradation priorities that are identified, whether by individual researchers, non-governmental organizations or global fora, to ensure that they are as valid in technical terms as they may be fashionable in political ones.

Policy Implications

Whether the impacts of land degradation are short-term or long-term, reversible or irreversible, human populations incur damage to their economic activity and social welfare. Factors other than land degradation may be involved, but, irrespective of the precise causality, the damage constitutes a political issue and requires an appropriate policy response.

Hitherto, environmental scientists have commonly investigated specific land degradations and sought to formulate appropriate preventative or mitigatory measures. Such measures aim to provide a rational basis for land-management decisions by politicians and planners. At times, the tendency has existed, especially on the part of mid-latitude scientists, to formulate technical options that are over-elaborate or otherwise unsuited to local land conditions or that poorly fit the social realities of the areas in question. In this respect, local soil conditions, property relations or cultural traditions can equally undermine a supposed technical solution to a specific degradation.

As far as the immediate land user is concerned, however, factors other than the availability of an appropriate technical solution affect the management of land degradation. Small farmers in the tropics may be so preoccupied with meeting subsistence needs that they cannot attend to longer-term measures that would protect or restore the quality of their land. Likewise, corporate land users, even if well aware of the long-term benefits of minimizing land degradation as a basis for sustainable exploitation, often opt for short-term profits at the expense of the environment in the course of, say, logging operations in Southeast Asia (Hurst, 1990) or cattle ranching in Amazonia (Fearnside, 1980; Eden *et al.*, 1990).

Where individual or corporate behaviour degrades the environment, more effective government involvement in land management is required. However, as at individual or corporate levels, governments have multiple and often conflicting priorities, which, in the tropics as elsewhere, commonly result in neglect of the environment. In recent decades, governments in the tropics have become more aware of the need for environmental protection.

Indeed, detailed legislative frameworks and dedicated government agencies are often in place which, if used effectively, would go a long way to resolving many of the environmental problems in both rural and urban–industrial areas. On occasions, governments are constrained by inadequate or misleading technical information, but more often they are deflected from, or simply ignore, their formal environmental obligations. Explicit priority may at times have to be given to resolving short-term economic or social crises at the expense of environmental resources. However, more culpable behaviour is often evident on the part of governments whose performances are characterized by inept or corrupt management that has little or no concern with land degradation or related issues.

Meanwhile, global awareness of the environment has increased in recent decades. External bodies ranging from international funding agencies and developed world governments to non-governmental organizations and the world media have variously proceeded to monitor, advise, criticize and even coerce governments in the tropics in respect of their environmental performance. The governments in question rarely appreciate such attentions which they see as infringing their authority or threatening national sovereignty. Even so, the reality of contemporary global politics, particularly since the United Nations Conference on Environment and Development in 1992, is that environmental issues are increasingly global issues that transcend national boundaries. Intrusive as this may appear and incompatible as it often is with the ambitions of individual states, it is increasingly the context in which land degradation and other environmental issues are, for better or worse, being approached (Anon., 1993).

Discussion

Although land degradation has traditionally been an environmental or technical issue, its causes and effects are linked to socio-economic conditions and its prospects related to political frameworks. Even assuming that suitable technical options exist for managing land degradation, there is no guarantee that such options will be adopted, even if longer-term national or global interests are served thereby. In other words, neither individual, corporate nor governmental responsibility for the environment can be assumed, and any policies aimed at managing land degradation must recognize this.

In these circumstances, it is firstly desirable to ensure that any technical options for managing land degradation are the best that current scientific knowledge allows. Some degradations, like soil nutrient depletion, are relatively uncomplicated and locally manageable in technical terms. Others, like deforestation, are less straightforward on account of their broad bio-physical feedbacks and overall technical complexity. At times, exaggerated impacts and 'worst-case scenarios' have been proffered for these and other degradations, and have latterly been justified in terms of sounding the alarm about environmental deterioration (Lovejoy, 1989) or of 'advocacy with respect to the environment' (Heywood and Stuart, 1992). Additional author-ity has been conferred thereon by resorting to the 'precautionary principle' which expresses the desirability of risking taking too much action rather than too little in the face of perceived threats (Heywood and Stuart, 1992). The principle is a valid response to technical uncertainty, but, in practice, needs to amount to more than a spurious cover for environmental propaganda. Its most useful function would be to provoke sufficient research on land degradation and other environmental issues to render itself a redundant principle. Some progress is being made, but the technical basis for under-standing and managing land degradation is still inadequate.

Secondly, it needs to be recognized that managing land degradation, critical as it is, is but one of a number of priorities in tropical countries. Governments often accord higher priority to developing commercial agri-culture or mining operations, implementing land colonization or urban housing schemes, or establishing strategic road-building projects. In spite of periodic pressure from radical environmental groups for severe constraints on tropical land development, the political problems involved are immense, and, in most cases, the best that can be hoped for is an enhanced compromise between development and environmental protection (Pearce, 1989; Eden, 1990, 1994). In practice, this means the promotion of sustainable exploitation systems and proper conservation programmes and the rigorous adoption of environmental impact analysis wherever significant developments are pro-posed. The aim must be to ensure that the linkage between development and degradation is acknowledged by politicians and planners in the tropics, and that conflicting economic, social and environmental priorities become ex-plicit political issues.

Thirdly, it has to be recognized that, even where environmental issues are part of the political process, it is often difficult for governments to implement environmental policies, even if they wish to. This is particularly so in tropical countries where severe economic constraints already exist, over which the government has little immediate control. The constraints may be massive overseas debts that absorb much of the national revenue or adverse terms of trade that limit generation of income in the first place. Externally applied policies, like the structural adjustment programmes of the World Bank, also severely constrain a government's scope for action, even if those policies are narrowly beneficial (Zinyama, Chapter 12). Likewise, internal structural factors limit the scope for action. Designated agencies for environmental management may be in place, but, if trained staff are unavailable or a shortage of vehicles or roads exists, there is little prospect of implementing existing environmental policies. In other cases, government performance is limited by security problems involving civil disturbance, guerrilla activity or drug trafficking. Again the scope for effective land management is curtailed.

Finally, it must be recognized that the quality of government in the tropical world, as elsewhere, is imperfect and that rational environmental, as well as economic and social, objectives are often compromised by individual or collective ignorance, incompetence or corruption. The conventional wisdom of the West is that an educated and informed population and a more open, democratic system of government is some regulator of such ills. There is an understandable disinclination in parts of the tropics to adopt a Western political model, but in such circumstances an alternative system is required that adequately handles environmental as well as other issues. Authoritarian political systems, whether of the left or the right, have a poor record in this respect. Thus, the quality of governance in the tropics is of critical importance to land degradation and as influential as any technical factor. Making governments more accountable for their environmental policies and actions, via internal or external feedbacks, is part of the way forward, and one in which the role of electorates, non-governmental organizations and the media is critical.

Given these constraints, what are the prospects for managing land degradation in the tropics? Some grounds for optimism exist, but the situation is uncertain as it depends ultimately on the political response. Thus far, as many chapters in the volume testify, that response has often been inadequate. Nevertheless, some progress has been made in translating land degradation into a political issue, which is a prerequisite to its management. The progress has not been fast enough to alleviate the fears over land degradation, but, in comparison to the neglect of tropical environmental issues that existed prior to the 1970s, a momentum of environmental concern has been generated that offers some hope for the future. The momentum has developed over the years from local attention to specific degradations to a broader concern with more general issues, some of which have lately been addressed at the global political level, notably at the United Nations Conference on Environment and Development in Rio de Janeiro.

Admittedly, such fora have their limitations, as do many of the more permanent institutions whose areas of concern encompass the environment. But it is difficult to envisage other ways forward that offer a lasting prospect of managing land degradation in the tropics. Meanwhile, the performance of relevant institutions, whether funding agencies or non-governmental organizations, needs constantly to be monitored by individuals and groups of individuals at all levels in order to ensure that appropriate environmental policies are adopted and that the momentum of concern is sustained.

References

Adams, W.M. (1990) *Green Development. Environment and Sustainability in the Third World*. London: Routledge.

Ahn, P.M. (1970) *West African Soils*. Oxford: Oxford University Press.

Anon. (1992) Map of land degradation/Mexico. *Newsletter, Critical Zones in Global Environmental Change*, **2**, 33–4.

Anon. (1993) Environmental protection or imperialism? *Nature*, **363**, 657–8.

Aubréville, A. (1949) *Climats, Forêts et Désertification de l'Afrique Tropicale*. Paris: Société d'Edition Géographiques, Maritimes et Coloniales.

Bahn, P. and Flenley, J. (1992) *Easter Island, Earth Island*. London: Thames and Hudson.

Barrow, C.J. (1991) *Land Degradation. Development and Breakdown of Terrestrial Environments*. Cambridge: Cambridge University Press.

Blaikie, P. and Brookfield, H. (1987) *Land Degradation and Society*. London: Methuen.

Bojo, J.P. (1991) Economics and land degradation. *Ambio*, **20**, 75–9.

Bordia, S. (1993) The mining industry of Papua New Guinea. In *Centenary Conference, Adelaide, 30th March–4th April 1993*. Australian Institute of Mining and Metallurgy Publication Series 2/93, Victoria, pp. 139–45.

Brown Jr, K.S. and Brown, G.G. (1992) Habitat alteration and species loss in Brazilian forests. In T.C. Whitmore and J.A. Sayer (eds), *Tropical Deforestation and Species Extinction*. London: Chapman and Hall, pp. 119–42.

Brown, M.J.F. (1974) A development consequence – disposal of mining waste on Bougainville, Papua New Guinea. *Geoforum*, **18**, 19–27.

Chisholm, A. and Dumsday, R. (1987) *Land Degradation: Problems and Policies*. Cambridge: Cambridge University Press.

Dasmann, R.F., Milton, J.P. and Freeman, P.H. (1973) *Ecological Principles for Economic Development*. London: John Wiley and Sons.

Diamond, J.M. (1991) Twilight of Hawaiian birds. *Nature*, **353**, 505–6.

Dickson, D. (1993) Academies urge population control. *Nature*, **365**, 382.

Eden, M.J. (1974) The origin and status of savanna and grassland in southern Papua. *Transactions of Institute of British Geographers*, **63**, 97–110.

Eden, M.J. (1990) *Ecology and Land Management in Amazonia*. London: Belhaven.

Eden, M.J. (1994) Environment, politics and Amazonian deforestation. *Land Use Policy*, **11**, 55–66.

Eden, M.J., McGregor, D.F.M. and Vieira, N.A.Q. (1990) Pasture development on cleared forest land in northern Amazonia. *Geographical Journal*, **156**, 283–96.

Falesi, I.C. (1976) Ecossistema de pastagem cultivada na Amazônia Brasileira. *Boletim Técnico, EMBRAPA/CPATU, Belém*, **1**, 1–193.

Fearnside, P.M. (1980) The effects of cattle pasture on soil fertility in the Brazilian Amazon: consequences for beef production sustainability. *Tropical Ecology*, **21**, 125–37.

Gilbert, A. (1974) *Latin American Development. A Geographical Perspective*. Harmondsworth: Penguin Books.

Gómez-Pompa, A., Vázquez-Yanes, C. and Guevara, S. (1972) The tropical rain forest: a non-renewable resource. *Science*, **177**, 762–5.

Goodland, R.J.A. and Irwin, H.S. (1975) *Amazon Jungle: Green Hell to Red Desert?* Amsterdam: Elsevier.

Gourou, P. (1961) *The Tropical World. Its Social and Economic Conditions and Its Future Status.* London: Longman, Green & Company.

Grainger, A. (1990) *The Threatening Desert. Controlling Desertification.* London: Earthscan Publications.

Hall, A.L. (1991) *Developing Amazonia. Deforestation and Social Conflict in Brazil's Carajas Programme.* Manchester: Manchester University Press.

Hamilton, D.C., Taylor, D. and Vogel, J.C. (1986) Early forest clearance and environmental degradation in south-west Uganda. *Nature*, **329**, 164–7.

Hasan, S. and Mulamoottil, G. (1994) Natural-resource management in Bangladesh. *Ambio*, **23**, 141–5.

Hayter, T. and Watson, C. (1985) *Aid. Rhetoric and Reality.* London: Pluto Press.

Helldén, U. (1991) Desertification – time for an assessment. *Ambio*, **20**, 372–83.

Heywood, V.H. and Stuart, S.N. (1992) Species extinctions in tropical forests. In T.C. Whitmore and J.A. Sayer (eds), *Tropical Deforestation and Species Extinction.* London: Chapman and Hall, pp. 91–117.

Hulme, M. (1989) Is environmental degradation causing drought in the Sahel? An assessment from recent empirical research. *Geography*, **74**, 38–46.

Hurst, P. (1990) *Rainforest Politics. Ecological Destruction in South-East Asia.* London: Zed Books.

INPE (1992) Deforestation in Brazilian Amazonia. *Revista SELPER*, Santiago, **8** (2), 35–6.

Ives, J. (1991) Floods in Bangladesh: who is to blame? *New Scientist*, **130** (1764), 34–7.

Johnson, D.L. and Lewis, L.A. (1995) *Land Degradation: Creation and Destruction.* Cambridge: Blackwell.

Kasperson, R. and Heffernan, J. (1992) Proposed IGU Commission on critical situations and regions in global environmental change. *Newsletter, Critical Zones in Global Environmental Change*, **2**, 2–4.

Kummer, D.M. (1992) Remote sensing and tropical deforestation: a cautionary note from the Philippines. *Photogrammetric Engineering & Remote Sensing*, **58**, 1469–71.

Lean, J. and Warrilow, D.A. (1989) Simulation of the regional climatic impact of Amazonian deforestation. *Nature*, **342**, 411–13.

Lovejoy, T. (1989) The obligations of a biologist. *Conservation Biology*, **3**, 329–30.

Manshard, W. (1974) *Tropical Agriculture. A Geographical Introduction and Appraisal.* London: Longman.

Martini, P.R. (1992) Panamazonia Project: an executive report. *Revista SELPER*, Santiago, **8** (2), 26–30.

Myers, N. (1992) Population/environment linkages: discontinuities ahead. *Ambio*, **21**, 116–18.

Nugent, S. (1990) *Big Mouth. The Amazon Speaks.* London: Fourth Estate.

Nunn, P.D. (1990) Recent environmental changes on Pacific islands. *Geographical Journal*, **156**, 125–40.

Nye, P.H. and Greenland, D.J. (1960) *The Soil under Shifting Cultivation.* Farnham Royal: Commonwealth Agricultural Bureaux.

O'Hara, S.L., Street-Perrott, F.A. and Burt, T.P. (1993) Accelerated soil erosion around a Mexican highland lake caused by prehispanic agriculture. *Nature*, **362**, 48–51.

Ohlin, G. (1992) The population concern. *Ambio*, **21**, 6–9.

Parry, J.T. (1986) Background, perspective, and issues for remote sensing in the tropics. In M.J. Eden and J.T. Parry (eds), *Remote Sensing and Tropical Land Management*. Chichester: John Wiley and Sons, pp. 337–60.

Pearce, F. (1989) Kill or cure? Remedies for the rainforest. *New Scientist*, **123** (1682), 40–3.

Prance, G.T. and Schubart, H.O.R. (1978) Notes on the vegetation of Amazonia. 1. A preliminary note on the origin of the open white sand campinas of the lower Rio Negro. *Brittonia*, **30**, 60–3.

Prowse, M. (1994) Towards a leaner, greener bank. *Financial Times*, London, 11 Jul.

Redclift, M. (1987) *Sustainable Development. Exploring the Contradictions*. London: Methuen.

Reiner, E.J. and Robbins, R.G. (1964) The middle Sepik plains, New Guinea. A physiographic study. *Geographical Review*, **54**, 20–44.

Rich, B. (1994) *Mortgaging the Earth. The World Bank, Environmental Impoverishment and the Crisis of Development*. London: Earthscan Publications.

Richards, P. (1985) *Indigenous Agricultural Revolution. Ecology and Food Production in West Africa*. London: Hutchinson.

Richards, P.W. (1973) The tropical rain forest. *Scientific American*, **229** (6), 58–67.

Richey, J.E., Nobre, C. and Deser, C. (1989) Amazon river discharge and climate variability: 1903 to 1985. *Science*, **246**, 101–3.

Sadik, N. (1994) Dr Sadik's remarks at PrepCom press conference. *Newsletter of International Conference on Population and Development*, Cairo, **14**, 5–6.

Schreckenberg, K., Hadley, M. and Dyer, M.I. (eds) (1990) *Management and Restoration of Human-impacted Resources. Approaches to Ecosystem Rehabilitation*. MAB Digest 5. Paris: United Nations Educational, Scientific and Cultural Organization.

Serrão, E.A.S., Falesi, I.C., de Veiga, J.B. and Neto, J.F.T. (1979) Productivity of cultivated pastures on low fertility soils in the Amazon of Brazil. In P.A. Sanchez and L.E. Tergas (eds), *Pasture Production in Acid Soils of the Tropics*. Cali: Centro Internacional de Agricultura Tropical, pp. 195–225.

Shankland, A. (1993) Brazil's BR-364 Highway. A road to nowhere? *The Ecologist*, **23** (4), 141–7.

Shukla, J., Nobre, C. and Sellers, P. (1990) Amazonian deforestation and climate change. *Science*, **247**, 1322–5.

Skole, D. and Tucker, C. (1993) Tropical deforestation and habitat fragmentation in the Amazon: satellite data from 1978 to 1988. *Science*, **260**, 1905–10.

Steadman, D.W. and Kirch, P.V. (1990) Prehistoric extinction of birds on Mangaia, Cook Islands, Polynesia. *Proceedings of National Academy of Sciences, USA*, **87**, 9605–9.

Steadman, D.W., Pregill, G.K. and Olson, S.L. (1984) Fossil vertebrates from Antigua, Lesser Antilles: evidence for late Holocene human-caused extinctions in the West Indies. *Proceedings of National Academy of Sciences, USA*, **81**, 4448–51.

Townshend, J., Justice, C., Wei, L., Gurney, C. and McManus, J. (1991) Global land cover classification by remote sensing: present capabilities and future possibilities. *Remote Sensing of Environment*, **35**, 243–55.

Tucker, C.J., Dregne, H.E. and Newcomb, W.W. (1991) Expansion and contraction of the Sahara desert from 1980 to 1990. *Science*, **253**, 299–301.

Verstraete, M.M. (1986) Defining desertification: a review. *Climatic Change*, **9**, 5–18.

Watters, R.F. (1971) *Shifting Cultivation in Latin America*. Rome: Food and Agriculture Organization of the United Nations.

Wooding, M. and Attema, E. (1994) *South American Radar Experiment. SAREX-92. Workshop Proceedings, 6–8th December 1993, ESA Headquarters, Paris*. Paris: European Space Agency.

World Bank Group (1994) *Learning from the Past, Embracing the Future*. Washington DC: World Bank Group.

World Commission on Environment and Development (1987) *Our Common Future*. Oxford: Oxford University Press.

Young, M. and Ishwaran, N. (1989) *Human Investment and Resource Use. A New Research Orientation at the Environment/Economics Interface*. MAB Digest 2. Paris: United Nations Educational, Scientific and Cultural Organization.

2

Radar Remote Sensing: Land Degradation and Hazard Monitoring

John T. Parry

Concern for environmental issues has increased substantially since the Stockholm Conference on the Environment convened by the United Nations Organization in 1972. The United Nations Conference on Environment and Development in Rio de Janeiro in 1992, although beset with political wrangling, has increased awareness of the extent of environmental degradation and addressed the formidable task of creating a political framework in which such issues as biodiversity, pollution control and sustainable development can be seriously considered. A future in which economic growth no longer depends on the exploitation of natural resources requires a comprehensive environmental data base, the political will to conserve and rehabilitate rather than abuse the environment and considerable managerial skills. The task is formidable, particularly for developing countries which carry the burden of large and growing populations, unemployment, poverty, haphazard growth of infrastructure, depletion of resources and increasingly serious environmental degradation.

It is fortunate that at a time when the possibility of an imminent environmental crisis has been appreciated, technology can provide synoptic imaging and data gathering, and the computing capacity to manage the information. Remote sensing and geomatics are at the forefront of this technology. However, experience demonstrates that developments and innovation in remote sensing, data handling and environmental management far outpace the capacity to make effective use of these powerful tools. The 'absorption capacity' of a user group is limited, first, by its awareness of the availability of a particular technology, and, second, by its experience of analysis and potential applications. The 'application capacity' of a user group is often limited by the inadequate infrastructure in a country or region. The essential link is that of education and training, not only technical and professional skills in handling and applying new techniques, but also the education of planners and policy-makers with regard to the potential of the available technology. The goal of remote sensing in the 1990s must be that of increasing the absorption and the application capacities in those areas where environmental problems are most severe.

Cloud conditions severely limit conventional aerial photography in much of the tropical zone. Multispectral satellite systems like Landsat and SPOT have proven operational value for a variety of applications in less cloudy areas of the world, including parts of the tropics (Soulsby, Chapter 8; Nagaraja and Gautam, Chapter 11), but their use is again limited in more humid tropical areas. Landsats 1, 2 and 3 carried on-board tape recorders. This allowed collection of an archive of imagery covering most of the earth's surface, which served as a prime data source for earth scientists, agronomists and economists. The archive is now somewhat dated, and, unfortunately, Landsats 4 and 5 do not have a recording capability. Thus, imagery capture is now limited to areas within range of receiving stations, with the result that no recent coverage exists for significant sections of east Asia and parts of Africa.

It is only with radar remote sensing, involving active systems at microwave frequencies, that reliable data collection and effective imaging can be guaranteed for the tropics. In this chapter, attention focuses on the special features of radar as a data source in monitoring environmental degradation and damage and in orchestrating planning responses.

Special Capabilities of Side-Looking Airborne Radar (SLAR)

Perhaps the most important feature of radar is its day/night, all-weather capability. Radars are active systems operating at frequencies of 8 to 40 Ghz. The wavelength is large relative to the size of water droplets, allowing the penetration of haze, cloud, fog and even rain. Radar is the only viable sensing system for cloud-persistent equatorial areas (Trevett, 1986). It is there that extensive SLAR coverage has been obtained in the last 30 years. There are also other areas in the tropics and subtropics with relatively few good days (cloud-free, haze-free, etc.) for the flying season, where SLAR operations can provide valuable environmental data.

A second important capability of radar is the provision of synoptic coverage of the terrain with broad swath widths which allow complete and continuous-strip coverage of linear features, such as river valleys and coastlines with their hinterlands (Figure 2.1). Mosaics can provide block coverage of large geographical areas, and landscape patterns are portrayed at a level of detail superior to that achieved in reconnaissance field surveys.

A third capability of airborne radar systems is rapid data acquisition. For example, the ERIM system with a swath width of 18 km operating in a turbo-prop Convair 580 with a ground speed of 550 km hr^{-1} can image 10 000 km^2 hr^{-1}. The Aeroservice SAR operating in a Caravelle jet during the *Projeto Radambrasil* (1973–79) with a swath width of 37 km and a ground speed of 550 km hr^{-1} achieved a mapping rate of 31 450 km^2 hr^{-1}.

Airborne radar has additional capabilities as a function of the look angle. In all SLAR systems, the energy pulse 'illuminates' the ground at relatively large incidence angles. This is analogous to a low sun angle in aerial photography. Radar shadowing emphasizes small differences in elevation, creating landscape texture enhancement. Changes in the shape, frequency and pattern of numerous individual shadows call attention to significant changes in landscape texture. These may be associated with geological, geomorphological, vegetative or land use variations.

In general, the specular rebound of microwave energy is directed away from the receiver since few smooth surfaces in the terrain are oriented perpendicular to the incident beam. However, in corner reflector configurations the specular rebound is returned to the receiver. Such configurations are quite common: for example, a steep river bank and the adjacent water surface, or the side of a building and the adjacent street or parking area. Even a clear-cut line in a forest canopy can give a characteristic bright signature matched by the radar shadow on the near side of the clearing.

The effects of roughness or surface micro-relief are less obvious. Many natural surfaces have small-scale features comparable in amplitude to radar wavelengths. The radar response can be summarized as follows: surfaces composed of elements having an amplitude half the wavelength of the radar will behave as specular reflectors, whereas surfaces composed of elements significantly larger than the radar wavelength will behave as diffuse reflectors and generate backscatter. The sensitivity of airborne radar to slope variation (macro-relief) and roughness (micro-relief) provides a capability in differentiating rock types and in terrain analysis that is unique.

This capability is limited in humid areas with closed-canopy, arboreal vegetation, since the frequencies of most airborne radar systems are too high to allow penetration of multiple leaf layers. In such situations, the radar illuminates the canopy surface, and roughness is a function of crown texture, i.e. branch geometry and leaf size, configuration and spacing. Comparisons of canopy contours with those of the subjacent ground surface in forested terrain show that the canopy topography is a systematically subdued version of the latter.

The final feature of airborne radar systems that needs mention is the stereoscopic capability. It is perfectly possible to lay out flight paths that

Area – South central Panama
System – Westinghouse Ka-band (HH)

A   the Americas bridge
B   Albrook air base
C   causeway
F   fishing vessels
G   Gamboa City
H   Howard airport
M   Miraflores lock
P   Panama City
R   Lago Chagres (water-level regulation for the canal)
S   range scale
T   timing marks (azimuth scale)
V   shipping in the canal
Y   shipping at anchor
Z   heavy precipitation

Figure 2.1 SLAR imagery of the Panama canal. (Reproduced courtesy of Goodyear Aerospace Corporation, Arizona Division.)

obtain appropriate base–height ratios for the parallax required in stereo-scopic imagery (Figure 2.2). However, radar stereoscopy differs from the conventional stereoscopy of aerial photography in that radar is a slant-range presentation even when corrected to give the appearance of a ground-range (vertical) view; thus the effects of range and shadow remain. In spite of these limitations, interpretation from the radar stereo view is improved and the confidence level of interpreted data increased (Koopmans, 1974).

Relatively little stereo radar imagery has been obtained. However, a pseudo-stereoscopic effect is created in all SLAR imagery as a result of the oblique illumination. This effect is particularly valuable in portraying relief (Figure 2.1). It results from the fact that microwave energy does not experi-ence atmospheric scattering, so there is no radar illumination within shadow areas. The sharp boundary between illuminated and non-illuminated areas produces a high-contrast image compared with the more gradual tonal transitions of aerial photography.

Environmental Degradation and Radar Remote Sensing

Each new environmental report seems to add to the litany of hazards and debilitating effects resulting from the increased intensity of human impacts. The countries most prone to these effects are generally tropical or sub-tropical, and the difficulties are compounded because many governments have only a limited response capability, the result of low GNP, poor infra-structure, and political weakness or indifference. There is a geographical face to all these problems. They involve climatic zones, mosaics of rock, soil, vegetation, crops and urban communities. To understand the scale and global teleconnections of these problems it is essential to have access to accurate and timely information. The keys to understanding environmental degradation are the synoptic viewing of regional patterns and change de-tection.

Radar remote sensing has much to offer in this regard, yet has been grossly underestimated as a monitoring tool. Set against the successes of the SLAR surveys of the late 1960s and early 1970s in Central and South America, Indonesia and New Guinea was the relatively high cost and the sense that the operations were experimental with little prospect of repeat coverage. A lack of experience in interpreting SLAR imagery, coupled with limited incentives for exploring the potential of radar remote sensing, accounted for a diminish-ing interest once the initial surveys were completed. The economic climate of the 1980s dampened this interest still further, apart from a brief flurry of activity stimulated by the SEASAT and Shuttle (SIR-A and -B) projects. The launch of three operational radar satellites, namely, ERS-1 in 1991, JERS-1 in

Area – Esmeralda,
Departamento
Atabapo, Estado
Amazonas,
Venezuela
System – Goodyear
(GEMS) X-band
(HH)
Stereo – same-side
viewing geometry

+   location mark
(lat. 3°10′N long.
65°30′W)

D   Cerro Duida
(800–1200 m) part
of the western
margin of the
Guyana Shield

E   Rancho grande
Esmeralda

G   areas of savanna
burned to
improve grazing

R   areas of denser
grass and shrub
cover

S   radar signal
distortion (range
streaking)

Figure 2.2 Stereoscopic SLAR imagery of savanna in southern Venezuela (for viewing with a pocket stereoscope). (Reproduced courtesy of Goodyear Aerospace Corporation, Arizona Division and International Aero Service Corporation, Division of Litton industries.)

1993 and the Canadian RADARSAT in 1995, has renewed interest in microwave systems. Is the user community ready for these developments? Probably not. Many environmental specialists are unfamiliar with the radar image and radar systems. In consequence, they are reluctant to consider microwave remote sensing as a valid tool in the investigation of environmental problems. In order to increase awareness of the value of radar data in examining environmental issues, a series of interpretative examples are offered in the pages that follow. Each addresses an important aspect of environmental degradation and management. They are not intended as final or exhaustive studies, but as exemplars and stimulants in the wider applications of radar remote sensing.

Flooding – Inundation

The increased incidence of flooding in many river systems is a direct result of mismanagement of the watershed through deforestation or overgrazing. There are serious economic consequences not only in terms of flood damage *per se*, but also with regard to infrastructure renewal. Bridges, spillways, culverts and levees built according to design criteria based on 50-year-old hydrological data are generally inadequate to cope with the increased volatility of the present-day river regimes.

Radar is an excellent tool for flood assessment because of the format of the imagery – a continuous swath that can follow the river course, and the capability for real-time monitoring at peak flood even in adverse weather conditions. Acquisition of flood data using airborne visible or infra-red techniques is often limited or prevented by the cloud and rain associated with the flooding. Radar's day–night, all-weather capability is particularly valuable for monitoring the dynamics of flooding following the breaching of levees or control structures and for co-ordinating emergency operations (Ormsby *et al.*, 1985).

The catastrophic floods in the Mississippi valley in July 1993 provided a salutary reminder that developed nations are not immune to such hazards. Figure 2.3 demonstrates the value of satellite radar imagery for operational evaluation of flood situations. Image A (ERS-1, 31 May 1993) shows normal conditions at the Mississippi–Missouri confluence. Image B (ERS-1) shows the flood limits on 14 July 1993. The colour composite (Plate 1) shows conditions at the peak of the flood (2 August 1993) at the Illinois–Mississippi confluence just west of the area in images A and B. A SPOT XS scene (18 July 1988) was used as the base for the colour composite image and this was overlaid with hydrological detail (channel outline and flood limits) from multi-date ERS-1 images (31 May and 2 August 1993).

ERS-1 provided information about water stage conditions throughout the flood episode and within 24 hours of acquisition these data were available to

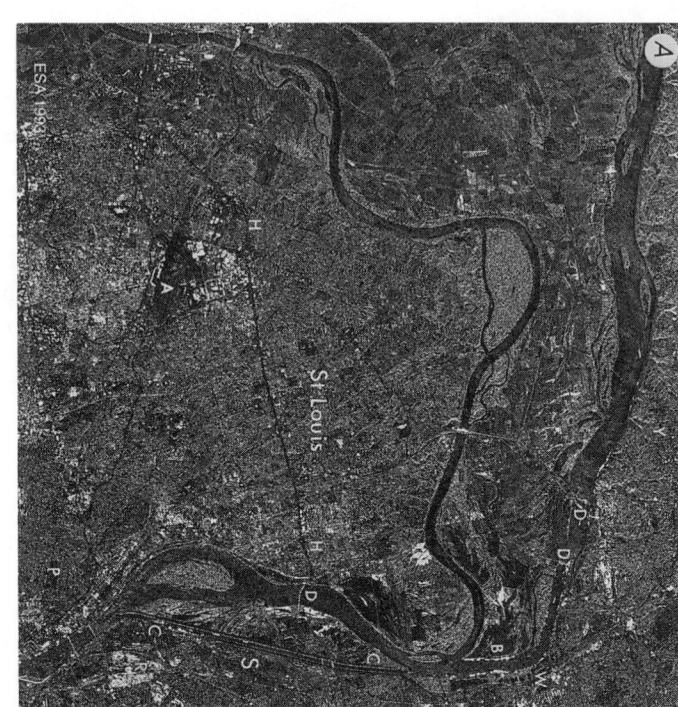

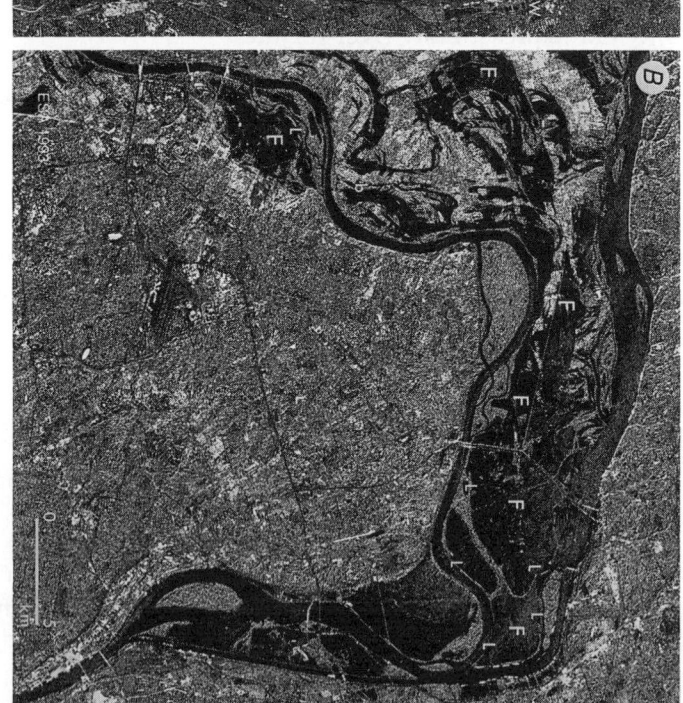

Figure 2.3 European Radar Satellite (ERS–1) images of pre-flood (31 May 1993) and flood conditions (14 July 1993) at the Mississippi–Missouri confluence. (Reproduced courtesy of ESA for ERS–1.)

Area – northeast Missouri–western Illinois, USA
System – ERS-1 C-band (HH)
A Lambert – Saint Louis airport
B barge train

C Chain of Rocks canal	**P** fairground park
D control dam	**R** rail yards
F Flooded cropland	**S** East Saint Louis
H Interstate 270	**W** Wood River (fuel tanks)
L levee	**Y** forest and scrub

the agencies engaged in control and emergency operations. Cloud conditions prevented imaging of the area by conventional sensors during four days at the most critical period. ERS-1 imagery of the inundated areas showed that the July 1993 Mississippi flood peak reached above the 100-year limits and in some areas covered parts of the 500-year hazard zone.

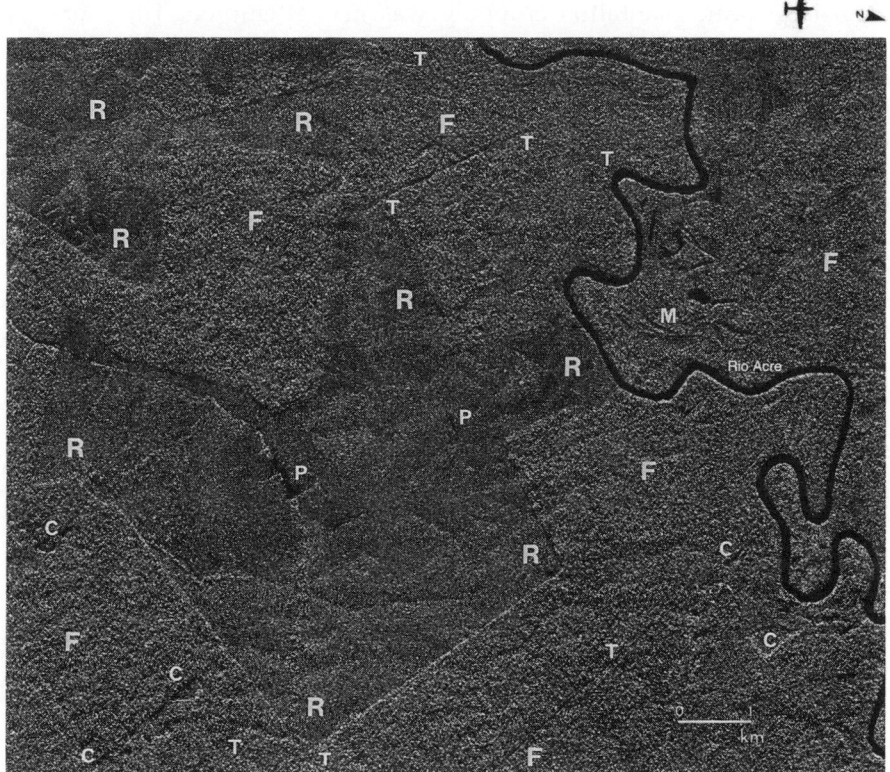

Figure 2.4 SLAR imagery of rain forest clearance for ranching in Brazil. (Reproduced courtesy of RADARSAT International. Data acquisition sponsored by CCRS, CIDA and ESA with support from INPE.)

Area – Acre Territorio Federal, Brazil
System – CCRS SAR-580 C-band (HH)
C clearings-shifting cultivation and settlements (rubber tappers)
F forest
M meander scrolls and backswamp
P stock ponds
R areas cleared for ranching
T trails

Forest Depletion – Ranching

Tropical deforestation has received increasing attention from the scientific community in the last 30 years. The local effects of forest removal and rapid reduction in fertility have been known since the beginning of the century when large-scale plantation cropping was first attempted. The more far-reaching effects on global biosphere teleconnections are only beginning to be understood. Analysis of these processes requires a knowledge of the extent of deforestation and reliable information on the rates of change of both continuing forest depletion and regeneration (Grainger, 1984). The same types of data obtained at prescribed monitoring intervals are also of great significance in regional and national planning.

SLAR and satellite radar systems (SRS) are eminently suited for this task. The texture (roughness) differences between the forest canopy and cleared areas are readily identifiable using visual interpretation methods, and there is good evidence to show that stages of regeneration can be identified using digital techniques (Riom and Le Toan, 1981; Wu, 1984). Radar returns from the edges of clear-cuts show some variability depending on orientation, incidence angle and the regularity of the felling. However, as shown in Figure 2.4, clear evidence of deforestation is provided by shadow alignment and corner reflector enhancement along the forest edge. These permit accurate area measurements. In this example from Acre in western Brazil, the area cleared for ranching in the centre of the image covers more than 30 km^2. Variations in the radar backscatter from parts of the cleared area indicate that regrowth is taking place.

This image is a RADARSAT simulation with an incidence angle of approximately 40° and a resolution of 10 m comparable to the satellite's narrow beam operating as described in a later section.

Forest Depletion – Timber Cutting

The logging of tropical hardwoods is one of the most destructive types of exploitation because, although only selected species are cut, there is damage to up to a third of the area by bulldozers and skidders in the search for and removal of these trees. Timber concessions are granted to large companies over extended periods of time and there is minimal control or even knowledge of their activities in remote areas of the equatorial rain forest.

A section of the coastal lowland in southwest Cameroon is shown in Figure 2.5. This image obtained by the Space Shuttle Imaging Radar (C-band HH) covers an area of approximately 1225 km^2 in département Kribi, east of the town of Campo. Since this is one of the more accessible sectors of the Cameroon, extensive logging has occurred. The relatively large cut-over area

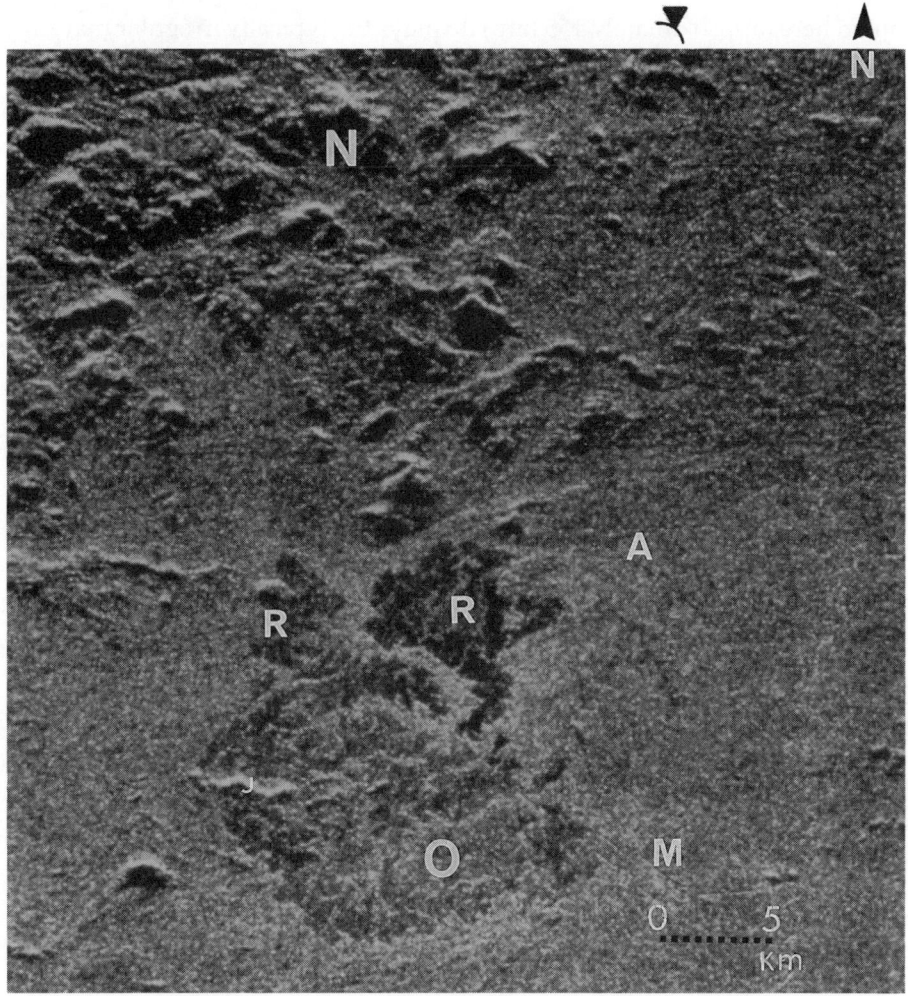

Figure 2.5 Shuttle Imaging Radar (SIR–A) imagery of logging operations in Cameroon. (Reproduced courtesy of the World Data Center 'A', Washington DC and Jet Propulsion Laboratory, Pasadena, California.)

Area – département Kribi, southwest Cameroon
System – SIR-A C-band (HH)
A   access roads to new logging areas
M   main logging road
N   Ngovayanqqui Mountains
O   older logged area
R   recently logged area (1981)

shown here (approximately 155 km^2) displays the typically irregular margins of logging activity and good contrast between recent (R) and earlier (O) operations. Hill outliers of the Ngovayanqqui crystalline massif within the cut-over remain forest-covered, as do some of the valleys, because of the difficulties of logging the steeper terrain. The change in radar backscatter between the cut-over and the untouched areas is explained by differences in surface roughness. Cut-over areas are relatively smooth compared to the irregularities of the forest canopy and, therefore, image in darker tones.

So far little research has been attempted on the feasibility of using radar remote sensing to obtain a measure of the rate of forest depletion in the tropics as a result of commercial timber harvesting operations (Stone *et al.*, 1989). Equally important is the assessment of regeneration. Satellite radar systems can supply accurate data on the location and extent of forest depletion and on long-term regrowth using the revisit capability. Such information is important for developing countries where basic resource data are limited and forest management is minimal.

Coastal Wetlands – Hazards and Management

Coastal zones contribute significantly to the economies of most countries. This is particularly true in the tropics where coastal wetlands provide a nutrient base for aquaculture and near-shore fisheries, sandy shores offer a potential for tourism, and estuarine and deltaic lowlands provide suitable soils and drainage conditions for rice cultivation. With varied and frequently intense human impact on the coastal zone, there is an increasing need for information about environmental conditions, changing land use, development conflicts and the like, in order to establish effective management planning (Bacon *et al.*, 1988).

Part of the Pacific coast of Costa Rica is shown in Figure 2.6. The step-faulted Fila Alta marks the northeast edge of the Valle del Diquis, a graben largely infilled by the Rio Terraba. The margins of its paleo-delta are indicated on the inset map. Comparison with the radar image shows good correspondence between the geomorphology and the land use: the extensive banana plantations of the United Fruit Company occupy the higher parts of the paleo-delta, the rice project is set in the backswamp areas and an aquaculture development is sited within the mangrove. Heavy sedimentation and a gentle off-shore gradient have promoted the development of a broad mangrove belt which provides excellent protection against tsunamis and storm surges from the Pacific. However, the redistribution of present-day sediments by a northerly long-shore drift is apparent in the radar image. SLAR permits quick and accurate mapping of coastline detail, and comparison of this image with the 1:200 000 topographic map (1978 revision) shows

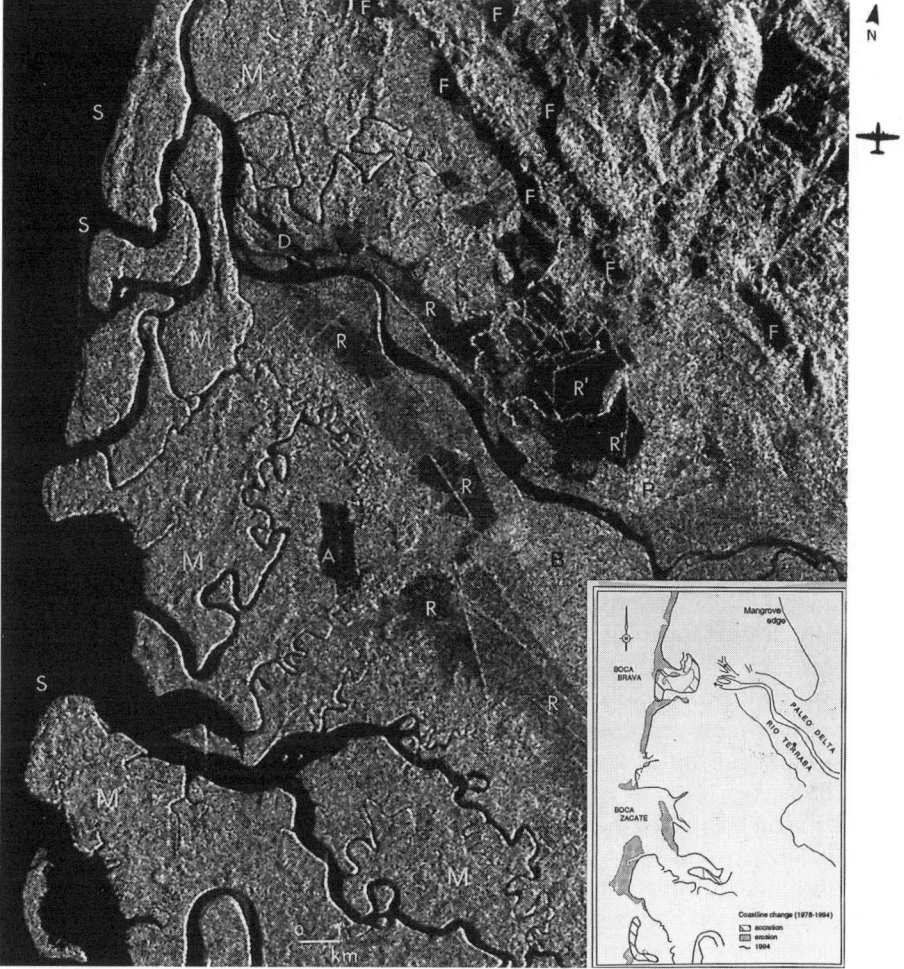

Figure 2.6 SLAR imagery of the mangrove coast of the Rio Terraba, Costa Rica. (Reproduced with permission of RADARSAT International. Data acquisition sponsored by CCRS, IDRC and ESA with support from the Instituto Geografico Nacional de Costa Rica – IGN.)

Area – Valle del Diquis, Provincia Puntarenas, Costa Rica
System – CCRS SAR-580 C-band (HH) wide mode (Radarsat simulation – standard mode 25 m resolution)

A aquaculture ponds	**R'** rice (flooded for planting)
B banana plantation	**D** distributory channels – paleo-delta
M mangrove	**F** Fila Alta – step faults
P Puerto Cortés	**S** mud flats – long-shore drift
R' rice stubble	

significant shoreline retreat. Changes that have occurred in the 16-year period (1978–94) are plotted on the inset map (Figure 2.6).

Mining Impacts and Mine Waste

By its very nature as an extractive industry, mining is destructive. Open pits leave a scar in the landscape and shafts contribute their spoil heaps, but these are minor impacts compared with the devastation wrought by area strip mining for coal. This is practised in coal basins with relatively thin over-burden. The area is cleared, drilled and blasted, and the overburden is removed and deposited in long, narrow banks which parallel the working face. As the seam is worked, the spoil banks succeed each other in regular series (Figure 2.7A).

Various environmental problems are associated with strip mining – erosion of the bare or thinly vegetated spoil banks and the discharge of highly mineralized, acidic water from the concentration of iron sulphides in the waste materials. In active mining districts, the area of stripping expands daily and can extend over tens of kilometres in a few months. Field mapping is time-consuming and there is generally no reason for updating the aerial photography of mining districts. Radar systems, particularly spacecraft systems, can provide initial surveys and repetitive monitoring of strip mining in a timely and cost-effective manner. In addition, the effectiveness of reclamation can be assessed and the potential sources of acidified water identified.

The Jharia basin in Bihar, shown in Figure 2.7B, is one of the most heavily industrialized areas in India. The gentle inward dip of the strata has allowed extensive strip mining of the Barakar coal beds (unit 3 in Figure 7B). The steep slopes of the spoil banks generate a strong radar return. Area measurements of these bright patches in the image show that more than 3500 ha have been affected between the Jamunia river and Dhanbad. These areas remain as unrestored spoil banks.

Special Capabilities of Satellite Radar Systems (SRS)

With the launch of SEASAT in 1978, spacecraft radar made its debut. This led to renewed interest in microwave remote sensing. The special features of satellite imaging radar were only fully realized after the launch of the Shuttle Imaging Radar (SIR-A) in the second Space Shuttle mission in November 1981 (Ford *et al.*, 1983).

The launch of the European Radar Satellite (ERS-1) and the Russian ALMAZ satellite in 1991 provided a renewed C-band (HH) and a new S-band

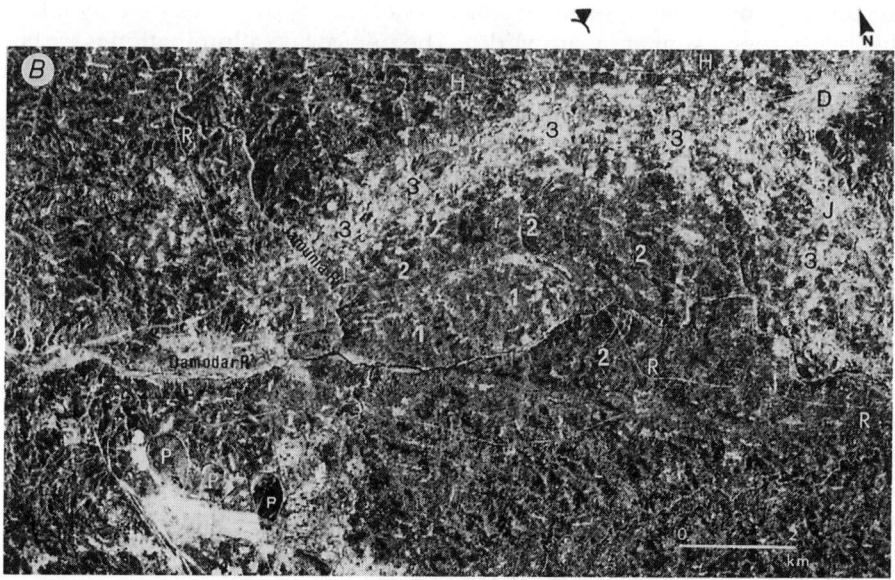

Figure 2.7A Detail of area strip mining for coal; and **B** European Radar Satellite (ERS–1) imagery of land disturbance resulting from strip mining in India. (Reproduced courtesy of the National Remote Sensing Agency, Department of Space, Government of India and ESA.)

Area – Topchanchi and Jharia districts, Bihar, India
System – ERS-1 C-band (VV)
Carboniferous succession in the Jharia basin:

4	Talchir Formation	I	industrial complex
3	Barakar Coal Beds (strip mining)	H	highway
2	Barren Measures	J	Jharia
1	Raniganj Formation	P	lagoon – industrial waste
D	Dhanbad	R	rail line

(HH) capability. With the launch of the Japanese SAR satellite (JERS-1) in 1992, an L-band system was added that had 25-m resolution over a 75-km swath. The decade of the 1990s is the first in which operational, multi-frequency radar imagery of the earth's surface is available from spacecraft.

Three special features of satellite imaging radars result directly from their orbital characteristics (Elachi, 1988). Spacecraft systems operate at altitudes of between 30 and 100 times greater than aircraft systems and this has an important effect on the imaging geometry. Airborne systems can only achieve a wide swath width by sweeping the radar beam across the surface from near to far range with progressive change in the depression (incidence) angle. This has a direct effect on the image characteristics since backscatter, foreshortening and radar shadowing all vary as a function of incidence angle. In contrast, the viewing geometry of a spacecraft SAR varies by only a few degrees across a similar swath width. The microwave illumination is essentially uniform and variations in radar return due to incidence angle effects are minimal.

The second advantage of spacecraft radar systems is their freedom from the effects of atmospheric turbulence. Even when equipped with motion compensation devices, airborne imagery is often marred by motion artifacts resulting from pitch, roll and yaw. Spaceborne systems operate well above the atmosphere and are unaffected by erratic movements due to turbulence.

A third feature of spacecraft systems relates to the speed and frequency of data acquisition. With a satellite in near-polar orbit, the earth continues its rotation within the orbit resulting in consecutive imaging swaths. RADAR-SAT with an orbital period of 101 minutes has an effective ground speed of $26\,800$ km hr^{-1} compared with perhaps 550 km hr^{-1} for a jet aircraft. Not only is the rate of coverage faster, but the swath can be significantly wider than for aircraft systems. This is important when one remembers that approximately 19.5 million km^2 of the earth's land surface experience perennial cloud cover that can only be penetrated by radar systems.

RADARSAT – The New Technology of Spacecraft SAR

Experience with SEASAT and the Shuttle radar demonstrated that in the long term the most effective way of collecting radar data was by satellite. Regular revisits with repeat coverage of critical areas and processes are of great importance for countries in the developing world where there is an urgent need for information about resources and environmental problems. RADAR-SAT is an advanced SAR system with features designed to meet Canada's requirements for resource management and environmental monitoring – essentially the same needs as for large territories in the tropics. RADARSAT was launched in 1995 with a five-year operational life. A review of its capabilities provides an opportunity to examine the future role of satellite microwave remote sensing.

The RADARSAT SAR is a C-band (5.6 cm wavelength), horizontally polarized system. It will be launched into near-polar orbit (inclination 98.6°), permitting coverage of all the world's environmentally sensitive zones – polar ice-caps, desert margins, rain forests and ocean basins. RADARSAT has a large demand for power and so it requires lengthy periods of solar illumination. For this reason, a sun-synchronous, dawn–dusk orbit has been selected. This has important benefits because it means that the radar can operate on solar rather than battery power. Thus, there is no operational difference between ascending (south–north) and descending (north–south) passes, and nearly twice as many imaging opportunities are available.

A special feature of RADARSAT is the provision of a variety of imaging modes and a range of ground resolutions (Figure 2.8). These have been selected as a result of more than fifteen years' study of radar systems and user needs (Raney *et al.*, 1991). The standard image format provided by RADAR-SAT covers a 100-km swath positioned within an 'orbit accessibility zone' of 500 km spanning a range of incidence angles from 20° to 49°. This is achieved using a relatively large antenna (15 × 1.5 m) that provides variable beam directions and different elevation beam widths. Only one incidence angle is available at any one time, and the swath is continuous and parallel to the subsatellite track. At the maximum operation time of 28 minutes, this would result in the collection of 1.25 million km^2 of radar data. The provision of a range of seven incidence angles allows the user to select the most appropriate radar illumination angle for the task and for the terrain conditions.

In addition to the standard beam mode, RADARSAT has a wide-swath mode available at incidence angles between 20° and 39° and a fine-resolution mode offering five positions in the incidence angle range 37° to 48°. The size of the ground resolution cell using the standard beam is 25 × 28 m, comparable to Landsat TM, while the fine-beam mode offers 11 × 9 m resolution, comparable to SPOT. The combination of selectable incidence angles and selectable resolving power provides RADARSAT with great flexibility. For example, in a single orbit it would be possible to obtain high resolution information on the progress of a flash flood in southern Libya, to continue the routine monitoring of the shoreline of Lake Chad using the wide-swath mode, to identify illegal logging operations in the Cameroon, again at high resolution, and to map the extent of flooding in the coastal wetland of Gabon.

Ground coverage varies according to mode and latitude. In the ScanSAR wide-scan mode, RADARSAT offers 35 to 40 per cent coverage of tropical regions each day and complete coverage every four days. Thus, comparative coverage of tropical areas and rapid revisit become a reality.

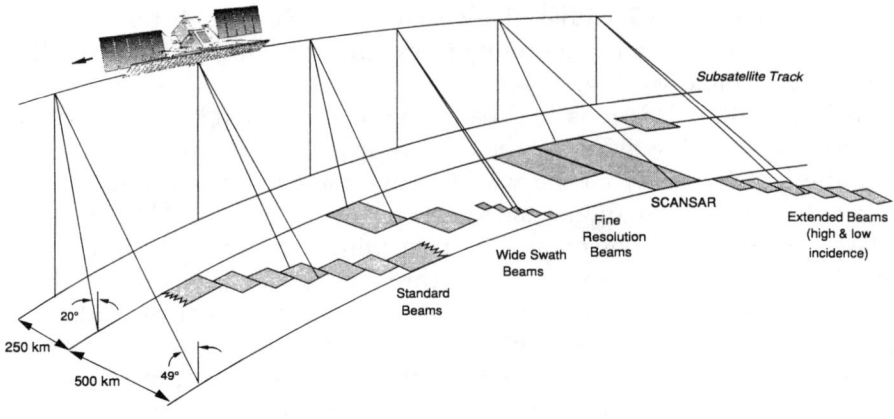

Beam mode	Swath width (km)	Resolution range x azimuth (m)	Incidence angles degrees (°)	Looks
Standard beams 7 positions >10% overlap	100	25 x 28	20 - 50	4
Wide swath beams 3 positions 3% overlap	150	25 x 28	20 - 40	4
Fine resolution beams 5 positions 10% overlap	50	11 x 9	37 - 48	1
Narrow ScanSAR	300	50 x 50	20 - 40	2
Wide ScanSAR	500	100 x 100	20 - 50	2
High incidence beams 6 positions 3% overlap	75	25 x 28	50 - 60	4
Low incidence beams	75	25 x 38	10 - 20	4

Figure 2.8 RADARSAT – imaging modes, swath widths and resolution.

RADARSAT will complement the European ERS-1 and Japanese JERS-1 satellites operating at C- and L-band wavelengths respectively. It will fill an important five-year window (1995–2000), one which may be critical for the monitoring and management of the earth's resources.

Conclusion

By the turn of the century, great advances will have been made in spacecraft radar technology. The real challenge is in their application: how radar data are used for resource survey, monitoring degrading environments and implementing management plans. It is thus appropriate to identify the main problems that must be addressed before effective use can be made of the new technology.

1 Our knowledge of the response of many targets to microwave energy is incomplete. Much remains to be learned about the effects of other factors, such as moisture content, wind, the micro- and macro-geometry of the individual elements of a land cover type, incidence angle, directional bias, etc. Knowledge is particularly scant with regard to C-band radars imaging tropical environments, hence the need for a co-ordinated programme of research.

2 Most resource scientists and managers are unfamiliar with radar remote sensing. Realizing the full potential of radar in land-degradation studies thus depends on the provision of training programmes in a broad field of applications. The aim is to develop an indigenous capability in a particular country in accordance with local needs.

3 The most appropriate procedure for transferring new technology is through pilot projects so as to assess the chances of success and develop modifications or alternate strategies. Before embarking on a full-scale radar project, it is vital to involve local users and potential users in a variety of disciplines. Land-use studies can serve as a bridge, since, by their nature, they are multi-disciplinary. An added benefit is that in many developing countries land-use data are a prerequisite for management planning.

4 Acquisition and processing of SAR data are a complex and costly process. They demand a high degree of technical expertise. Like many other aspects of technology transfer, the pace of development should be set by the host country in order to develop an indigenous capability and to amortize the cost (debt) over a period of time.

5 Radar is already of proven value in environmental studies, particularly those of a regional nature. Its role in more detailed studies has yet to be demonstrated. Such studies are urgently

needed to provide the touchstone against which the claims of the
marketing managers of different radar systems can be judged. It
would be foolhardy to promise results which are never realized, and
it would be short-sighted to leave unexplored potential avenues of
application. In this regard, there is no substitute for field-based
studies. The ultimate accuracy of any radar interpretation and
application hinges on the level of knowledge of what actually exists
on the ground.

References

Bacon, P.R., Deane, C.A. and Putney, A.D. (1988) *A Workbook of Practical Exercises in Coastal Zone Management for Tropical Islands*. London: Commonwealth Secretariat.

Elachi, C. (ed.) (1988) *Spaceborne Radar Remote Sensing: Applications and Techniques*. New York: Institute of Electrical and Electronics Engineers Press.

Ford, J.P., Cimono, J.B. and Elachi, C. (1983) *Space Shuttle Columbia Views the World with Imaging Radar: the SIR-A Experiment*. Pasadena: JPL Publication 82–95, Jet Propulsion Laboratory.

Grainger, A. (1984) Quantifying changes in forest cover in the humid tropics: overcoming current limitations. *Journal of World Forestry Resource Management*, **1**, 3–63.

Koopmans, B.N. (1974) Should stereo SLAR be preferred to single strip imagery for thematic mapping? *ITC Journal* 1974 **3**, 424–44.

Ormsby, J., Blanchard, B. and Blanchard, A. (1985) Detection of lowland flooding using active microwave systems. *Photogrammetric Engineering and Remote Sensing*, **51**, 317–28.

Raney, R.K., Luscombe, A.P., Langham, E.J. and Ahmed, S. (1991) Radarsat. *Proceedings of Institute of Electrical and Electronics Engineers*, **79**, 839–49.

Riom, J. and Le Toan, T. (1981) Relations entre des types de fôrets de pins maritimes et la rétrodiffusion radar en bande L. In *Proceedings of International Society of Photogrammetry International Colloquium on Spectral Signatures of Objects in Remote Sensing*, Avignon, pp. 455–66.

Stone, T.A., Woodwell, G.M. and Houghton, R.A. (1989) Tropical deforestation in Para, Brazil: analysis with Landsat and Shuttle Imaging Radar. In *International Geoscience and Remote Sensing Society '89 Symposium Proceedings*, Vancouver, Vol. 1, pp. 192–5.

Trevett, J.W. (1986) *Imaging Radar for Resource Surveys*. London: Chapman and Hall.

Wu, S.T. (1984) Analysis of synthetic aperture radar data acquired over a variety of land cover. In *Institute of Electrical and Electronics Engineers Transactions on Geoscience and Remote Sensing*, GE22, pp. 550–6.

PART II

Degradation in Tropical Forests

PART II

3

Forest Degradation in the Tropics: Environmental and Management Issues

Michael J. Eden

Widespread concern arose over tropical deforestation during the 1970s. P.W. Richards, one of the pioneers of ecological studies in tropical forests (Davis and Richards, 1933; Richards, 1952), was also one of the first to warn of the increasing forest destruction (Richards, 1973). Over the years, many other ecologists, geographers and anthropologists have likewise expressed concern (Gómez-Pompa et al., 1972; Denevan, 1973; Davis, 1977; Myers, 1984; Grainger, 1993a). As research on tropical forests has increased, it has become apparent that deforestation has many direct and indirect causes, and that numerous environmental and social impacts derive therefrom. This complexity similarly characterizes the newer and broader issue of forest degradation, which includes partial damage to the forest as well as complete deforestation.

Much uncertainty exists with regard to technical aspects of forest degradation and to their implications for forest policy and management. Current concerns include the limited data on the rate, total extent, and bio-physical impacts of deforestation, and the lack of any real perspective on forest degradation as a whole. In respect of forest policy and management, problems exist in seeking to achieve sustainable exploitation and explicit protection of forest land in lieu of degradation, especially in areas where small-scale colonization or large-scale commercial developments are rife. The social impact of forest degradation on traditional forest dwellers also causes concern. In this introductory chapter, initial attention is paid to deforestation, but the emphasis then shifts to the more general issue of forest degradation; the latter is examined in both environmental and management terms.

Environmental Aspects of Forest Degradation

Hitherto most attention has focused on the assessment of deforestation and its associated bio-physical feedbacks. Given that Landsat imagery has been available for more than two decades, it is disappointing that more reliable information on the total extent and rate of deforestation has not been assembled. Accurate satellite-derived data exist for some areas, notably Brazilian Amazonia which contains approximately a third of the remaining tropical forest (Grainger, 1993b; Eden, 1994). Several pan-tropical forest surveys based on Landsat or NOAA AVHRR imagery have been implemented, but a clear overall picture has yet to emerge (FAO, 1993; Malingreau, 1993). As Table 3.1 indicates, discrepancies exist between recent estimates of the extent and rate of deforestation in the tropics.

Table 3.1 Estimates of the extent and rate of clearance of tropical moist forest.

	Extent of tropical moist forest (million ha)	Rate of clearance (million ha yr^{-1})
Myers (1989)	800	14.2
FAO (1992)	1282	12.2

Sources: After Myers (1989), FAO (1992) and Grainger (1993b).

Uncertainty also prevails over the bio-physical feedbacks of deforestation. This is less of a problem in respect of local soil chemical or physical impacts, for which precise data are more readily obtained, but much uncertainty exists over the more complex, large-area impacts of deforestation. The limited primary data relating, for example, to loss of biodiversity or changing climatic or hydrologic regimes make assessments of large-area impacts uncertain. Conflicting and, at times, exaggerated predictions of the environmental impacts of deforestation have consequently been made (Brown and Brown, 1992).

While there are continuing attempts to improve the quality of data on deforestation and on related bio-physical feedbacks, it is increasingly recognized that the post-clearance land cover is also a significant environmental variable. In places, a tendency has existed to stress the 'non-renewability' or limited regenerative capacity of the tropical forest after clearance and to assume that the forest is often replaced by a herb-dominated, low-biomass cover. As Eden (Chapter 4) indicates, this assumption is questionable and in need of review not least because it impinges on predictions about the climatic and hydrologic feedbacks of deforestation.

It is also recognized that tropical deforestation, which has hitherto been the main focus of attention, is only part of the broader process of forest degradation. Deforestation, i.e. clear-cutting for agricultural, ranching or other purposes, continues in the tropics, but it frequently coexists with partial disturbance of or damage to the forest, typically as a result of selective logging (Grainger, 1993b; Eden, 1994). The latter degradation, which is not generally included in estimates of deforestation, also needs to be assessed in respect of both its local and regional impacts.

More explicit attention to the post-clearance land cover and to the partial disturbance of or damage to forests implies an increased concern with the status and dynamics of secondary vegetation, especially secondary forest. Hitherto, secondary communities have been relatively neglected in the tropics, but their extent is rapidly increasing with the scale of human disturbance (Chazdon, 1994). The classification and monitoring of such communities are an urgent requirement, although it represents a much greater challenge to both ground survey and satellite remote sensing than does simple deforestation monitoring.

Managing Forest Degradation

Technical uncertainties of the kind outlined above provide a poor basis for formulating, and persuading governments to implement, policies that incorporate sustainable exploitation and explicit protection as means of limiting forest degradation. In places, the direct conversion of forest land to other uses will inevitably occur, but parallel efforts are required to maintain large tracts of sustainably managed forest and of natural forest. An appropriate balance is needed that satisfies broader development objectives, but also provides adequate environmental protection and recognizes the needs of local forest dwellers. As a basis for this, further research is needed to clarify technical aspects of forest degradation.

Such clarification is required at both regional and local levels, and, in this section, examples are offered of the kind of applied research that can enhance forest management. At regional level, there needs to be an adequate management response to the anticipated feedbacks of large-area forest degradation. Hitherto, large-area deforestation has been the prime concern (Myers, 1984; Wilson, 1988), but, as indicated, a broader approach is now required. In this respect, Grainger's formulation (Chapter 5) of a taxonomy of degraded forest types in Southeast Asia, as a basis for mapping the distribution of degraded forest itself, is a starting point for assessing potential forest rehabilitation. The specific context is that of mitigating global climate change by sequestering atmospheric carbon, but the approach is equally relevant in assessing and managing other feedbacks of forest degradation.

At the local level, Greer *et al.* (Chapter 6) show how monitoring of erosion in a commercially logged forest in Malaysia provides a basis for water conservation and erosion control by forest managers. The techniques are equally applicable in other tropical regions where commercial logging is undertaken. Equivalent studies in areas of forest clearance for agricultural or ranching purposes offer similar benefits.

Even where appropriate technical data on tropical forests and forest degradation are available, their application to actual forest management is often limited, whether in the context of land development, conservation or restoration. This is very disappointing, given the legislative and institutional frameworks that often exist for forest management. Admittedly, in places, legislation can encourage land degradation, as in Brazilian Amazonia where tax incentives and other provisions have accelerated deforestation (Binswanger, 1991), but, in other contexts, existing legislation and institutions, if properly employed, can serve to promote sustainable and protective land uses and thus limit forest degradation.

In many countries, for example, National Parks and other reserved forest areas have been officially designated in recent years, and, on paper at least, provide an encouraging start to conserving forest biodiversity. In the late 1980s, an estimated 66 million ha of protected moist forest existed in the tropics, representing some 5 per cent of the historic forest cover (Simberloff, 1986; Sayer, 1991). Significant contributors to the total were Brazil (9.6 million ha) and Australia (8.1 million ha) (Eden, 1990; Sayer, 1991). However, not all 'paper parks' are adequately protected on the ground, with many having suffered damaging incursions by small farmers or logging companies (Tchamie, 1994). Indigenous reserves or other designated tribal lands in forest areas are also vulnerable to invasion by sundry land developers, with predictable consequences for the forest and its local inhabitants (Fearnside and Ferreira, 1984; Colchester, 1994).

In addition, attempts are commonly made by governments and their agencies to regulate land development in forest areas. In respect of logging activities, for example, regulations frequently exist regarding the minimum tree size or maximum timber volume that can be extracted from timber concessions (Repetto and Gillis, 1988; Hurst, 1990). In recent years, legislation has also been enacted in countries like Thailand, Indonesia, Philippines and Brazil that bans the export of raw logs. This is intended to encourage local processing of timber and to conserve remaining forests (Goodland *et al.*, 1990). Such regulations are well-intentioned, but are poorly respected on the ground.

Many reasons exist for the failure to regulate forest degradation, whether in the form of clearance or partial damage. Ineffective, day-to-day management by government and its agencies is an immediate and obvious explanation, given the relevant legislative and regulatory frameworks that exist in many countries. These frameworks are clearly required, but, as is evident, are no guarantee of effective forest management. This reflects the fact that many attempts to regulate damage by logging companies or clearance by small farmers or ranchers are addressing symptoms rather than the real causes of forest degradation (Vanclay, 1993). Thus, increasing population and growing human impoverishment in the tropics often drive people to exploit and degrade the forest irrespective of the legality or long-term sense of their actions. Elsewhere, forest degradation reflects the commercial opportunism of national or multinational companies that pursue short-term profits at the expense of the natural capital of the forest. The greed and corrupt practices of such companies are widely acknowledged (Hurst, 1990; Vanclay, 1993), but they also operate within political systems that are at times inherently corrupt, inefficient, and unconcerned about the environment.

At a fundamental level, it is thus the broader socio-political environment, tolerating and ultimately causing forest degradation, that precludes effective forest management. At the national level, many governments in the tropics accept or encourage forest degradation for reasons that range from the supposed national interest, through evident self-interest, to the fact that at times they simply cannot control it. Likewise, at international level, development agencies and their sponsoring governments promote so-called land development that may achieve specific economic benefits, but often causes unnecessary forest degradation. The 'greening' of these agencies, not least the World Bank, is supposedly occurring (Rich, 1994; World Bank Group, 1994), but a great deal needs to be done before their performance can be adjudged satisfactory. Greater accountability to electorates in the case of governments and to member governments in the case of international agencies is needed, if the appropriate balance between land development and environmental protection is to be achieved.

In this context, an important role exists for non-governmental organizations. In the developed world, they have acquired a significant influence in environmental affairs, but, in tropical countries, they are less influential and, at times, unacceptable to governments. Yet growing awareness of environmental issues will surely advance the cause of non-governmental organizations in the tropics, and allow them to bring greater pressure to bear on governments in the future (Smiet, 1993; Rich, 1994).

Conclusion

Forest degradation involves clearance of, or partial damage to, the existing forest cover. The effective management of forest degradation requires a knowledge of the status and dynamics of the forest system itself and of related post-clearance land covers. Forest degradation is also a political issue that involves national and international perceptions of the value of tropical forests as an economic, social and ecological resource. As indicated, there are multiple explanations for forest degradation in the tropics, and effective management demands a range of responses. They include the following:

1   The need to clarify the status and environmental impacts of forest degradation and to establish effective procedures for monitoring it.
2   The need to achieve both sustainable exploitation and explicit protection of the forest as a means of limiting forest degradation.
3   The need to make governments in the tropics and international agencies and their sponsors more accountable for their policies and actions so as to sustain effective forest management.

References

Binswanger, H.P. (1991) Brazilian policies that encourage deforestation in the Amazon. *World Development*, **19**, 821–9.
Brown Jr, K.S. and Brown, G.G. (1992) Habitat alteration and species loss in Brazilian forests. In T.C. Whitmore and J.A. Sayer (eds), *Tropical Deforestation and Species Extinctions*. London: Chapman and Hall, pp. 119–42.
Chazdon, R.L. (1994) The primary importance of secondary forests in the tropics. *Tropinet*, **5** (2).
Colchester, M. (1994) The new sultans. Asian loggers move in on Guyana's forests. *Ecologist*, **24**, 45–52.
Davis, S.H. (1977) *Victims of the Miracle. Development and the Indians of Brazil.* Cambridge: Cambridge University Press.
Davis, T.A.W. and Richards, P.W. (1933) The vegetation of Moraballi Creek, British Guiana: an ecological study of a limited area of tropical rain forest. Part 1. *Journal of Ecology*, **21**, 350–84.
Denevan, W.M. (1973) Development and the imminent demise of the Amazon rain forest. *Professional Geographer*, **25**, 130–5.
Eden, M.J. (1990) *Ecology and Land Management in Amazonia*. London: Belhaven.
Eden, M.J. (1994) Environment, politics and Amazonian deforestation. *Land Use Policy*, **11**, 55–66.
FAO (1992) The forest resources of the tropical zone by main ecological regions. Paper presented at UN Conference on Environment and Development, June, Rio de Janeiro.
FAO (1993) *Forest Resources Assessment 1990: Tropical Countries*. FAO Forestry Paper No. 112., Rome: United Nations Food and Agriculture Organization.
Fearnside, P.M. and Ferreira, G. de L. (1984) Roads in Rondonia: highway construction and the farce of unprotected forest reserves in Brazil's Amazonian forest. *Environmental Conservation*, **11**, 358–60.
Gómez-Pompa, A., Vázquez-Yanes, C. and Guevara, S. (1972) The tropical rain forest: a non-renewable resource. *Science*, **177**, 762–5.
Goodland, R.J.A., Asibey, E. O. A., Post, J.C. and Dyson, M.B. (1990) Tropical moist forest management: the urgency of transition to sustainability. *Environmental Conservation*, **17**, 303–18.
Grainger, A. (1993a) *Controlling Tropical Deforestation*. London: Earthscan Publications.

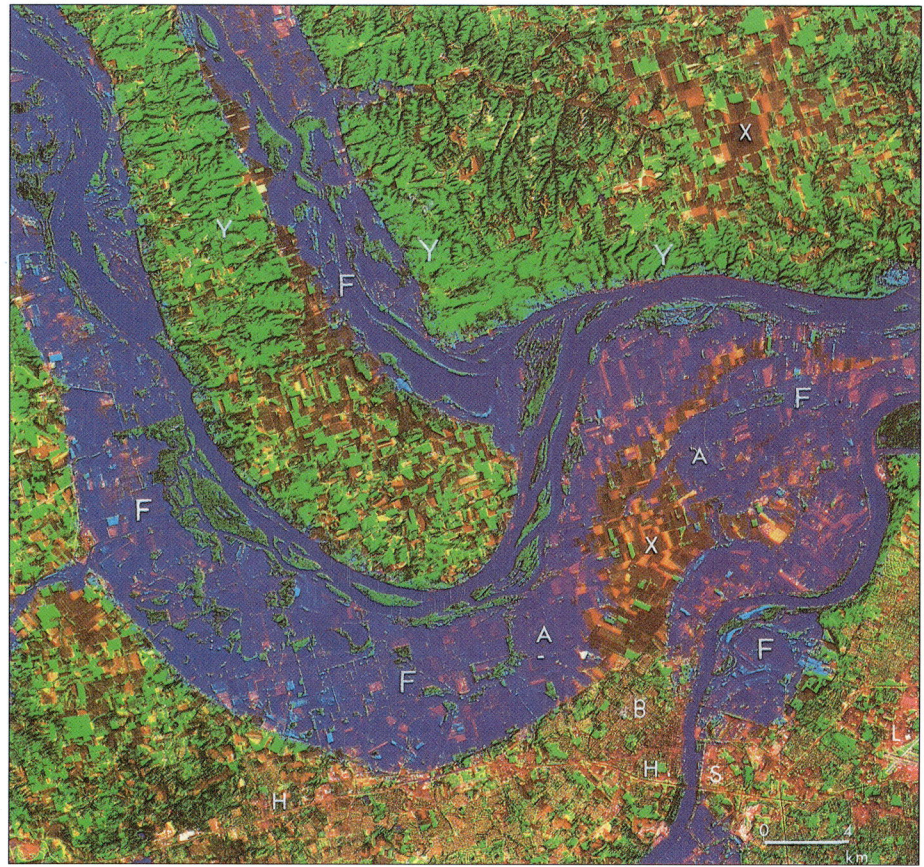

Area – northeast Missouri – western Illinois, U.S.A.
System – SPOT XS – ERS 1 C-band (HH) colour composite

A	airport (flooded)	**F**	flooded areas	**L**	Lambert–Saint Louis Airport	**X** cropland
B	Bridgton	**H**	Interstate 70	**S**	Saint Charles	**Y** forest and scrub

Plate 1 Flooding of the valley bottomland at the Illinois–Mississippi confluence. Composite image generated by merging SPOT multispectral data (pre-flood, 18 July 1988) with European Radar Satellite (ERS–1) data obtained on 31 May and 2 August 1993. (Reproduced courtesy of RADARSAT International. Processing by Space Remote Sensing Centre, Stennis Space Centre MS. SPOT satellite data made available by SPOT Image Corporation.)

 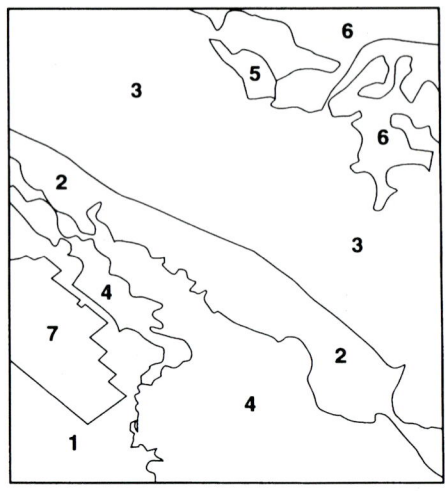

Manually Stretched
MSS False Colour Composite - December
Land Cover and Erosion Hazard:

1   Bare dry soils - high

2   Cultivated alluvial soils - moderate to high

3   Open canopy savanna woodland - low

4   Marshland - very low

5   Montane evergreen forest - very low

6   Tea plantation - very low

7   Irrigated sugar cane - very low

Plate 2 Manually stretched Landsat MSS false colour composite of bands 4, 5 and 7 for extract from scene 179/071 (22 December 1981) and associated land cover and erosion hazard, lower Shire valley, Malawi. (Digital data courtesy of NOAA.)

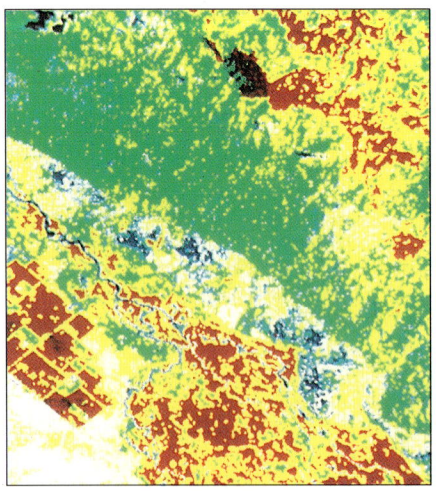

Parallelepiped Classification - December

White -   Bare dry soil

Blue -   Wet bare soil and burned vegetation

Green -   Medium cover vegetation

Yellow -   Moist bare soils with
low cover/ vigour vegetation

Red -   High cover / vigour vegetation

Plate 3 Parallelepiped classification of Landsat MSS bands 4, 5 and 7 for extract shown in Plate 2, lower Shire valley, Malawi. (Digital data courtesy of NOAA.)

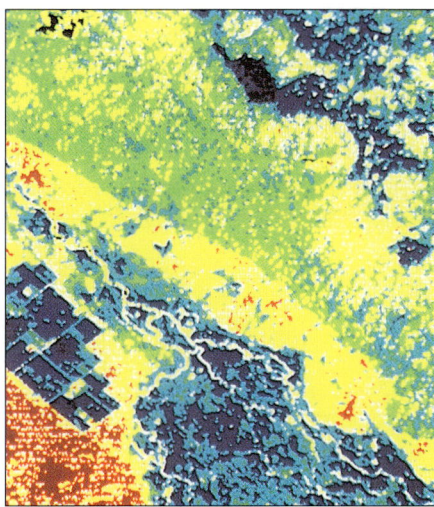

Erosion Hazard from
Classification

Red - High

Yellow - Moderate

Green - Low

Blue - Very low

Plate 4 Erosion hazard for lower Shire valley, Malawi, derived from parallelepiped classification shown in Plate 3. (Digital data courtesy of NOAA.)

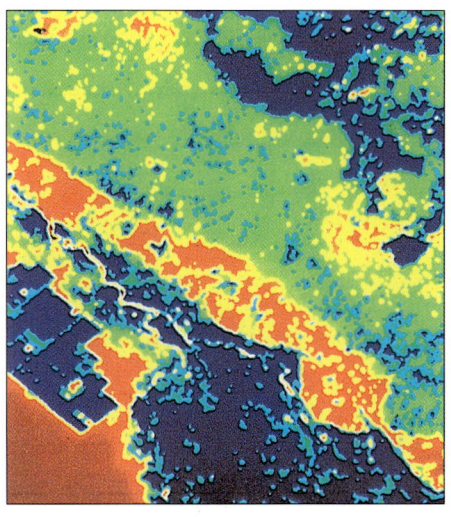

Near Infra-red / Red Vegetation Index
Erosion Hazard

Red - High

Yellow - Moderate

Green - Low

Blue - Very low

Plate 5 Erosion hazard for lower Shire valley, Malawi, derived from near infra-red/red vegetation index for extract shown in Plate 2. (Digital data courtesy of NOAA.)

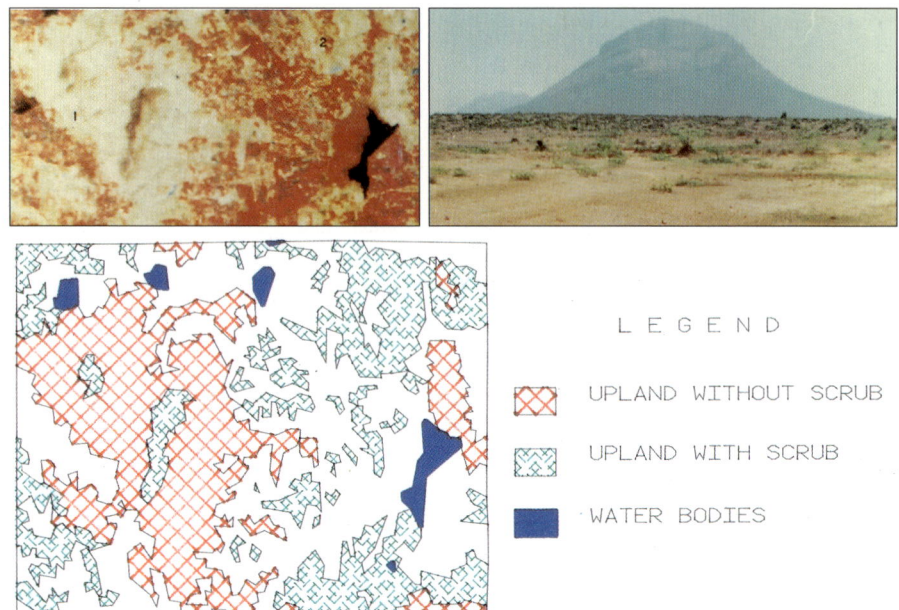

LEGEND

⬚ UPLAND WITHOUT SCRUB

⬚ UPLAND WITH SCRUB

◼ WATER BODIES

LANDSAT TM SIGNATURE BANDS 2,3,4 (blue, green, red)

Season:	Oct.-Nov. (dry season)
Terrain:	Variable lithology and slopes
Pattern:	Variable size units, irregular shape, scattered occurrence
Texture:	Fine to mottled
Colour:	Light to dark yellow to greenish blue varying with soil moisture and surface cover

Plate 6 Summary of field and Landsat image characteristics for Type 2 wasteland: upland without scrub in North Arcot, Tamil Nadu, India. (Courtesy of National Remote Sensing Agency, Hyderabad, India.)

Grainger, A. (1993b) Rates of deforestation in the humid tropics: estimates and measurements. *Geographical Journal*, **159**, 33–44.

Hurst, P. (1990) *Rainforest Politics. Ecological Destruction in South-East Asia.* London: Zed Books.

Malingreau, J-P. (1993) Satellite monitoring of the world's forests: a review. *Unasylva*, **44** (174), 31–8.

Myers, N. (1984) *The Primary Source. Tropical Forests and Our Future.* New York: W.W. Norton.

Myers, N. (1989) *Deforestation Rates in Tropical Forests and Their Climatic Implications.* London: Friends of the Earth (UK).

Repetto, R. and Gillis, M. (eds) (1988) *Public Policies and the Misuse of Forest Resources.* Cambridge: Cambridge University Press.

Rich, B. (1994) *Mortgaging the Earth. The World Bank, Environmental Impoverishment and the Crisis of Development.* London: Earthscan Publications.

Richards, P.W. (1952) *The Tropical Rain Forest. An Ecological Study.* Cambridge: Cambridge University Press.

Richards, P.W. (1973) The tropical rain forest. *Scientific American*, **229** (6), 58–67.

Sayer, J. (1991) Conservation and protection of tropical rain forests: the perspective of the World Conservation Union. *Unasylva*, **42** (166), 40–5.

Simberloff, D. (1986) Are we on the verge of mass extinction in tropical rain forests? In D.K. Elliott (ed.), *Dynamics of Extinction.* New York: John Wiley.

Smiet, A.C. (1993). Tropical forestry in the 21st century: limitations and opportunities. *Ambio*, **22**, 50–1.

Tchamie, T.T.K. (1994) Learning from local hostility to protected areas in Togo. *Unasylva*, **45** (176), 22–7.

Vanclay, J.K. (1993) Saving the tropical forest: needs and prognosis. *Ambio*, **22**, 225–31.

Wilson, E. O. (ed.) (1988) *Biodiversity.* Washington DC: National Academy Press.

World Bank Group (1994) *Learning from the Past, Embracing the Future.* Washington DC: World Bank Group.

4

Environmental Degradation and Forest Renewability in Amazonia

Michael J. Eden

Recurrent concern has been expressed in recent decades about forest clearance in Amazonia. Attention has commonly focused on the extent of clearance itself, the limited regenerative capacity of the forest, and the likely degradation of edaphic, hydrologic, climatic and biotic systems. In some respects, the concern is justified, but it has at times been accompanied by exaggerated and alarmist predictions and excessive emphasis on 'worst-case scenarios'. This has not been without benefit, as it has served to alert Latin American governments and the international community to the potential environmental degradation arising from large-area forest clearance. However, the current *mélange* of truth, prediction, speculation and propaganda provides an imperfect basis for formulating appropriate land-management policies in Amazonia.

The perspective adopted in this chapter is that although Amazonian forest clearance has commonly resulted in local environmental damage, it has not, as yet, had a disastrous regional impact in terms of predicted environmental feedbacks. This is mainly because the current extent of forest clearance remains relatively low, but it also reflects the fact that the forest itself is more resilient in the face of contemporary land colonization than has sometimes been assumed. The chapter thus examines the relationship between environmental degradation and forest renewability in Amazonia. Initially, the causes of forest clearance are outlined, and the nature and extent of anticipated environmental feedbacks are reviewed. Secondly, forest renewability and the post-clearance land cover are considered, both in respect of analogous forest/savanna boundaries and of areas of recent agricultural and pastoral clearance. Finally, the broader policy implications of the above are considered in relation to regional land management.

Causes of Forest Clearance

The causes of forest clearance in Amazonia are diverse and of varying importance across the region. Shifting cultivation and other small-scale farming activities cause widespread clearance, but coexist with large-scale commercial enterprises, notably cattle ranches and, to some extent, industrial plantations. Of late, extractive logging has also increased in parts of Amazonia, although its main effect is to damage rather than clear the forest. Other causes of forest clearance include road building, mining activities and hydro-electric developments.

Forest clearance has traditionally been associated with shifting cultivators, whose individual fields are mostly small and isolated in the forest. Some evidence exists that indigenous shifting cultivation has caused local deflection of forest to savanna (Scott, 1978; Huber *et al.*, 1984), but, for the most part, such cultivation, which has been practised for millennia, involves only temporary displacement of the forest.

Small-scale cultivation also occurs in recent colonization zones in Amazonia. In such areas, land settlement is often relatively concentrated and more extensive deforestation results. An early example was the Bragantina zone of eastern Pará in Brazil, where planned land colonization for agriculture, initiated in the late nineteenth century and based on 25-ha family holdings, resulted in extensive forest clearance that ultimately spread across some 3.0 million ha (Sioli, 1973). Similar planned colonization, based on 100-ha family plots, was initiated along parts of the Transamazon highway in central Pará in the early 1970s, and has likewise resulted in extensive deforestation. Initial land use in the area was agricultural, although some later development of larger-scale pastoral activities has occurred.

Large-scale commercial developments, particularly cattle ranching, have also been directly responsible for extensive forest clearance in Amazonia. Cattle ranching has been most widespread in Brazil, particularly in Pará and Mato Grosso States. Precise data on the area involved are lacking, but some 20 million ha have probably been cleared for pasture, mostly since the 1960s. The extent of individual holdings is highly variable, but many exceed 20 000 ha and some reach 100 000 ha and above. Similar clearance has occurred in parts of the Andean *oriente*, mainly in Colombia and Peru where 6 to 7 million ha of pasture have been so derived (Eden *et al.*, 1990). In these areas, the size of individual holdings is generally much smaller than in Brazilian Amazonia.

The development of commercial tree plantations has also contributed to forest clearance, albeit on a much smaller scale than cattle ranching. Familiar examples from Brazil include the Fordlandia and Belterra rubber plantations on the Tapajós river, dating from the late 1920s and early 1930s, and the Jari timber plantation in northern Pará, dating from the 1960s. *In toto*, these plantations were originally scheduled to cover some 3 million ha, but less than 20 per cent of that area has been cleared to date. Elsewhere in Amazonia, many smaller tree plantations have been established over the years, notably for rubber, oil palm, timber and cacao, but recurrent ecological and economic problems have limited their overall extent (Eden, 1994).

The third main impact on the forest is commercial logging, which, particularly in the lower Amazon region, has been practised for several centuries. However, logging activities have greatly increased since the mid-1970s, particularly in Brazilian Amazonia. Mechanized extraction has lately become more common, expanding apace with the developing road network (Eden, 1990). Even so, relatively low timber volumes (10 to 40 m^3 ha^{-1}) are extracted from most forests, with the result that localized canopy damage rather than complete forest clearance occurs. However, extraction rates are currently increasing (Uhl and Vieira, 1989), whilst the creation of logging trails through the forest provides ready access for ranchers and itinerant farmers, who establish themselves on the land and complete the process of clearance.

Environmental Impact of Forest Clearance

The amount of forest clearance in Amazonia has been the subject of uncertainty and dispute. An early estimate was that of Sommer (1976), who suggested that the extent of South American moist forest, largely but not exclusively Amazonian, had been reduced by some 37 per cent. Other earlier estimates were those of Myers (1979), who suggested about one-third, and Gentry and Lopez-Parodi (1980), who suggested that 20 to 25 per cent of the Amazonian forest had been cleared. Such estimates may approximate the extent of forest that has lately been subject to human disturbance, but they exaggerate actual clearance. A more recent estimate suggests that, by 1990, total Amazonian clearance amounted to some 65 to 70 million ha, which is 10 to 11 per cent of the historic forest cover; 41 to 42 million ha are in Brazilian Amazonia (INPE, 1992; Eden, 1994).

Although the local impact of forest clearance is often damaging, the effects at regional level have been less apparent. Even so, many commentators have envisaged positive feedbacks that are likely to degrade the regional environment as deforestation proceeds. Among the more alarming predictions has been Friedman's (1977, p. 7) assertion that Amazonian deforestation can be expected to result in 'extreme climatic effects' converting the region into 'a dry savannah similar to northeastern Brazil or the African Sahel'. He concluded that the forest would take thousands of years to regenerate. Comparable views were expressed by Gentry and Lopez-Parodi (1980), who envisaged extensive deforestation causing accelerated run-off and lower rainfall that might eventually convert much of Amazonia to 'near-desert' conditions.

In climatic terms, such predictions have been partially validated in that contemporary rainfall levels in Amazonia are generally acknowledged to depend on the efficient recycling of water by the forest cover, in whose absence rainfall would be expected to decline (Molion, 1976; Salati *et al.*, 1978). Such dependence is seemingly confirmed by computer simulations that show markedly lower evapotranspiration and rainfall levels when the Amazonian forest is replaced by a pasture cover (Lean and Warrilow, 1989; Shukla *et al.*, 1990; Nobre *et al.*, 1991).

Concern has also been expressed over the loss of biodiversity that would occur in the event of large-area deforestation. An early expression of this was Gómez-Pompa *et al.* (1972) who identified the danger of a mass extinction of forest species across the tropics. Similar views were expressed by Wilson (1988). More specifically, Barney (1982) suggested that a third or more of the species in Latin America's tropical forests would disappear by the year 2000, while Myers (1988, p. 32) has latterly suggested that 'a major extinction spasm in Amazonia is entirely possible'.

In reality there is, as yet, little, if any, direct evidence of the predicted macro-regional feedbacks of Amazonian deforestation (Sternberg, 1987; Eden, 1994). This is exemplified by Richey *et al.* (1989), who failed to detect any significant change in the discharge of the Amazon river attributable to recent changes in land-use patterns. Likewise, doubts have been expressed as to whether massive species extinctions are occurring, or are indeed imminent, in Amazonia or other tropical areas as a result of recent deforestation (Brown and Brown, 1992; Heywood and Stuart, 1992). In part, the lack of evidence reflects the limited extent of contemporary forest clearance, but it also relates to the response of the land itself to clearance. A critical consideration here, which is the concern of the present chapter, is the nature of the post-clearance land cover. The latter is important because whether cleared land develops a degraded, low-biomass herbaceous community or a regenerating woody cover affects the anticipated edaphic, hydrologic, climatic and biotic feedbacks.

In this respect, the assumption is periodically made that the forest is 'non-renewable' and thus likely to be replaced by a low-biomass cover of dry savanna (Friedman, 1977), pasture (Lean and Warrilow, 1989), or 'near-desert' (Gentry and Lopez-Parodi, 1980). In particular, Gómez-Pompa *et al.* (1972) explicitly described the tropical rain forest as a whole as 'a non-renewable resource' and argued that it was incapable across most of its extent of regenerating under current land-use practices. Particular concern was expressed over large-area clearance where, as a function of the reproductive characteristics of many primary forest species, natural forest regeneration was considered unlikely to occur.

Similar concern was expressed by Uhl (1982, 1988) in Amazonia. He argued that, although forest regeneration usually occurs after small-scale shifting cultivation, it is more problematic when larger areas are cleared and used intensively for pasture. In such cases, regeneration is constrained not only by the poor dispersal of tree seeds across larger cleared areas, but also by predation of the few seeds that do arrive and by the harsh micro-climatic conditions that confront them. Such views were reiterated by Nepstad *et al.* (1991), who additionally emphasized the role of fire in arresting forest regeneration by killing tree seedlings and saplings. Such fires frequently affect old pastures in seasonal forest areas of eastern Pará, and also extend into neighbouring logged forests where partial canopy opening has occurred and large amounts of combustible slash remain on the forest floor. Although Nepstad *et al.* (1991) acknowledge that forest clearance is 'not a terminal event in Amazonia', others, such as Lovelock and Kump (1994) persist in the alarmist view that the clearance of tropical forests is 'irreversible', at least on a timescale of human observation.

As indicated, the renewability or otherwise of the Amazonian forest cover is critical in evaluating the predicted environmental feedbacks of forest clearance. In places, a degraded low-biomass land cover may result from clearance, but it needs to be considered whether this represents a local 'worst-case scenario' or the general prospect for the region. In this respect, increasing evidence exists that forest regeneration commonly occurs on cleared forest land, and that the more alarmist views regarding the status of the post-clearance land cover are unjustified.

Forest Renewability and Post-clearance Land Cover

In spite of the availability of satellite imagery for Amazonia since the early 1970s, limited attempts have been made to monitor the post-clearance land cover as distinct from forest clearance itself. While the latter task is now more or less in hand through the efforts of the Brazilian *Instituto Nacional de Pesquisas Espaciais*, systematic information on the cover of cleared forest land is generally lacking. However, forest regeneration on such land is by no means precluded. In this respect, attention is firstly drawn to the analogous situation at forest/savanna boundaries in the region, and, secondly, data are presented on the land cover, and by implication regeneration processes, on cleared forest land.

Forest/Savanna Boundaries

Since Haffer's (1969) seminal paper on Amazonian Pleistocene forest refugia, considerable discussion of associated fluctuations in forest/savanna boundaries has occurred. Although some doubts have arisen regarding the validity of the refugia model (Endler, 1982; Colinvaux, 1987), recent palynological data (Absy *et al.*, 1991) reinforce earlier biogeographical and other evidence evoked in its support (e.g. Turner, 1971; Prance, 1973; Eden, 1974). Curiously, however, discussion of the refugia model has scarcely involved any attention to the ability of the forest to recover from its periodic, large-area contractions. It seems simply to have been assumed that the forest was renewable, albeit at a rate so slow as to be irrelevant to any discussion of contemporary forest management. While that may or may not be the case, the past record certainly confirms that no absolute constraint exists on large-area forest regeneration on 'cleared' forest land.

Admittedly, it is apparent from contemporary studies that recurrent savanna fires commonly preclude forest regeneration into adjacent savanna, but, if fire is controlled, relatively rapid local forest regeneration occurs. This has been demonstrated under experimental conditions at Calabozo in the Venezuelan *llanos* where, over a 20-year period, numerous forest and savanna trees have invaded an area of open-wooded savanna protected from fire (Medina and Silva, 1990). Similar regeneration has been reported at Emas in the Brazilian *cerrado*, where local control of burning since the 1960s has allowed the development of a forest cover in what was previously open-wooded savanna (Coutinho, 1982).

Local evidence of forest regeneration into savanna is also evident within Amazonia itself. At Maracá, on the western margin of the Rio Branco savanna in northern Brazil, savanna trees that were encountered within the margins of the flanking forest have been taken to indicate an extension of that forest in recent years (Milliken and Ratter, 1989; Furley and Ratter, 1990). In the adjacent southern Rupununi of Guyana, similar local forest advance into the savanna has occurred (Hills, 1964; Eden, 1986). In both instances, the process is associated with a reduction or cessation of savanna fires, related in the case of Maracá to the designation of an ecological reserve and in the case of the southern Rupununi to local withdrawal of Amerindian settlement and shifting cultivation (Milliken and Ratter, 1989; Eden and McGregor, 1992).

Such findings are not only of interest in terms of post-refugial rates of reforestation, but also suggest that while fire may be a critical constraint, such factors as seed predation and a harsh micro-climate do not preclude forest regeneration on 'cleared' land. Equally, the seemingly rapid rate of such reforestation raises questions about the extent to which seed dispersal constrains large-area forest regeneration. In this context, it is noted that existing savannas in the vicinity of Maracá and the southern Rupununi are frequently interpenetrated by riparian forest and contain scattered 'bush islands' that presumably serve as local seed sources across the larger savanna.

In summary, it appears that forest may be 'non-renewable' in the face of recurrent savanna fires, but that its regenerative capacity is not otherwise notably impaired, as suggested by its prompt extension when little or no burning occurs. The idea of neotropical savannas as a 'fire climax' is of course by no means new (Rawitscher, 1948), but it offers a useful perspective on the dynamics of regeneration on recently cleared land.

Recent Forest Clearance

Few empirical data exist on the general post-clearance land cover in Amazonia. In places, cleared land has reportedly been so degraded that forest regeneration is seriously impaired. Such land includes heavily used pastures along the Belém-Brasilia highway in eastern Pará, where land preparation, after initial cutting and burning, has depended on bulldozers to scrape off all vegetation and woody debris and on mechanical discing and levelling of the land before seeding with grasses (Uhl *et al.*, 1988). On abandonment, pastures of this kind only slowly accumulate above-ground biomass and rarely develop much woody regrowth. However, such sites are relatively rare, comprising less than 10 per cent of abandoned pastures (Uhl *et al.*, 1988).

Degraded conditions have also been reported from the Bragantina zone of eastern Pará, where early agricultural colonization of forest land occurred. Precise data on the current land cover are lacking, but the area was earlier reported to have little vegetation beyond scrub and brush growth (Egler, 1961 in Myers, 1991); the area has recently been described as one 'where there has never been extensive regrowth of the natural vegetation' (Nobre *et al.*, 1991, p. 962). These are valid enough observations, but it must be noted that, by Amazonian standards, the Bragantina zone is densely populated, often attaining 30 to 90 people km^{-2}, and is still widely cultivated on a temporary and, in places, permanent basis (Eden, 1990; Smith, 1992). Commercial agriculture, including tree cropping, is locally established, and the 'degraded' status of the land cover is more a reflection of continuing land occupance and land use than any inherent inability to regenerate to forest after what has misleadingly been described as 'a short-lived effort at agricultural settlement early this century' (Myers, 1991, p. 18).

In contrast, significant levels of forest regeneration are reported on cleared forest land in parts of Brazilian Amazonia. These include the majority of pastures along the Belém–Brasilia highway, which are typically abandoned after 6 to 12 years and thereafter develop secondary woody regrowth (Fearnside, 1979; Uhl *et al.*, 1988). Similar conditions are described by Lucas *et al.* (1993) in northern Amazonas State, where extensive forest clearance for crops and pasture occurred as a result of the opening of the Manaus–Boa Vista highway in the late 1970s. From examination of Landsat imagery, it is apparent that, by 1991, 'much of the previously cleared land north of Manaus supported secondary forest' (Lucas *et al.*, 1993, p. 3062). Likewise, in the vicinity of Boa Vista itself, Dargie and Furley (1994) report that large areas of cleared forest land, visible on 1978 Landsat imagery, had reverted to forest by 1985.

Most of the scattered information on post-clearance land cover in Amazonia is qualitative in nature, but more detailed studies are being initiated, again using sequential satellite imagery. One such is that of Mausel *et al.* (1993), who have examined the land cover along the Transamazon highway near Altamira, where extensive clearance for crops and pasture has been undertaken by colonists since the early 1970s. Within a study area of 267 000 ha astride the highway, Landsat images reveal that, by 1991, approximately 100 000 ha of forest had been cleared. Of this area, 13.5 per cent remained under cultivation, but the rest consisted of secondary regrowth, classified as 'initial succession' (29.2 per cent), 'intermediate succession' (41.4 per cent), and 'advanced succession' (15.9 per cent). Both the intermediate and advanced categories showed 'multicanopy' development with trees reaching heights of 8 to 12 m and more than 20 m respectively (Mausel *et al.*, 1993). This constitutes a significant level of forest regeneration.

Discussion

In spite of recurrent emphasis on the constraints to forest regeneration on cleared land (Gómez-Pompa *et al.*, 1972; Uhl, 1988; Nepstad *et al.*, 1991), a significant amount of regeneration is occurring in parts of Amazonia rather than general deflection to a low-biomass, herb-dominated cover. Under conditions of intensive or persisting land use, forest regeneration may be precluded, but a more characteristic sequence is one of initial 'slash and burn', short-term exploitation with minimal industrial inputs, and early land abandonment in the face of declining soil fertility and weed invasion, leading to forest regeneration. In these circumstances, no absolute constraint is offered by edaphic, micro-climatic or biotic factors. As is evident in adjacent savannas, fire is potentially limiting to forest regeneration, but even its impact is evidently partial, particularly where less combustible woody weeds promptly re-establish on the land. Moreover, as Nepstad *et al.* (1991) imply, fire is a manageable variable in the region.

The other significant factor reportedly affecting regeneration on cleared forest land is the scale of clearance and its effect on tree seed dispersal (Gómez-Pompa *et al.*, 1972; Uhl, 1988). Again field data indicate that large-area clearance may not be as limiting as suggested. This is evident, for example, along the Transamazon highway near Altamira, where many forest fragments persist within the general colonization zone (Mausel *et al.*, 1993), presumably providing multiple seed sources at least capable of promoting a secondary forest cover. Equally, in the very largest cattle ranches in parts of northern Mato Grosso, where blocks of 100 km^2 or more of forest have been cleared, it is evident from Landsat imagery that many forest fragments, notably of riparian forest, persist as local seed sources (Eden, 1990). The situation effectively replicates that found in the *llanos* and other savannas, where the herb-dominated cover is locally interspersed with riparian and other forest fragments.

The widespread regeneration of secondary forest on cleared land has implications for the predicted environmental feedbacks of clearance. Admittedly, a regenerating forest will initially be of lower stature and biomass than a primary forest and its growth rate will be limited in accordance with the intensity of previous exploitation (Nepstad *et al.*, 1991), but it will replicate the protective function of the primary forest to a greater extent than any herb-dominated cover. The detailed performance of secondary forest in this respect is largely unexplored, but it is surmised that accumulating leaf litter in secondary forest will progressively minimize surface water run-off and soil erosion (Hamilton, 1991), developing root systems will increasingly recycle soil nutrients and soil water (Eden and McGregor, 1992; Nepstad *et al.*, 1991), and a growing forest canopy will steadily increase evaporative flux to the atmosphere (Salati *et al.*, 1986). Likewise, the carbon store of a secondary forest will increase as regeneration proceeds. These processes will minimize the predicted edaphic, hydrologic and climatic feedbacks of forest clearance to an extent that will be lacking in a low-biomass herb-dominated cover.

In respect of predicted biotic feedbacks, particularly loss of biodiversity, the situation is less clear. Even if secondary forest commonly regenerates, large-area clearance may still threaten the survival of some primary forest species whose status is less secure since their seeds tend to be large, rather short-lived and dependent on animal dispersal (Gómez-Pompa *et al.*, 1972; Kubitzki, 1985). Even so, such species are often widely distributed across Amazonia (Heywood and Stuart, 1992), and, assuming that some large forest fragments survive as National Parks and the like, the species are not necessarily destined for extinction. Nevertheless, it is wise to pursue forest conservation policies that minimize the risk of biotic degradation.

Against this background, there is a need for continuing investigation and monitoring of the extent and distribution of regenerating secondary and remnant primary forest, both of which contribute to regulating the environmental feedbacks of clearance. Further investigation is also required of fire, which may serve to maintain derived herb-dominated communities, or, if effectively regulated, allow general forest regeneration. Fire is important since its broad susceptibility to management signifies that socio-political factors, as well as ecological processes, control forest regeneration. Some relevant conservation policies have been formulated in recent years, ranging from schemes to establish networks of National Parks and other reserves (Prance, 1990; Foresta, 1991) to more recent plans, particularly in Brazil, to regulate dry season burning (Nepstad *et al.*, 1991). Even so, in spite of continuing environmental propaganda and elaboration of a derivative 'precautionary principle' (Heywood and Stuart, 1992), only modest progress is being made with conservation policies in the face of widespread developmental pressures.

In summary, enhanced land-management strategies are required that will gain the support of developmental interests in Amazonia, without at the same time causing too much environmental degradation. This is not the place to explore the utility of land-use options like extractive reserves, mixed-forest silviculture, agroforestry and the like, but it is arguable that such options, coexisting with networks of National Parks, will help to accommodate governmental expectations of regional development as well as encourage a more positive commitment to balanced forest management. Such will be facilitated by the apparent natural tendency of the forest to regenerate on much cleared land, which in turn will limit the degrading feedbacks of clearance. In spite of recurrent alarms, Amazonia is not yet a 'red desert' (Goodland and Irwin, 1975) or a 'wasteland' (Friedman, 1977). Most of the region is still under forest, and, in spite of persisting developmental pressures, it is still possible to avoid gross environmental degradation, provided enhanced management strategies are devised and implemented as a step beyond the predictions and propaganda, alarmist or otherwise, of recent decades.

References

Absy, M.L., Cleef, A., Fournier, M., Martin, L., Servant, M., Sifeddine, A., Ferreira da Silva, M., Soubies, F., Suguio, K., Turcq, B. and van der Hammen, T. (1991) Mise en évidence de quatre phases d'ouvertures de la forêt dense dans le sud-est de l'Amazonie au cours des 60000 dernières années. Première comparaison avec d'autres régions tropicales. *Comptes Rendues Académie de Sciences*, Paris, 312, Série 2, 673–8.

Barney, G.O. (1982) *The Global 2000 Report to the President. Entering the Twenty-first Century*. Harmondsworth: Penguin Books.

Brown Jr, K.S. and Brown, G.G. (1992) Habitat alteration and species loss in Brazilian forests. In T.C. Whitmore and J.A. Sayer (eds), *Tropical Deforestation and Species Extinctions*. London: Chapman and Hall, pp. 119–42.

Colinvaux, P. (1987) Amazon diversity in light of the paleoecological record. *Quaternary Science Reviews*, **6**, 93–114.

Coutinho, L.M. (1982) Ecological effect of fire in Brazilian cerrado. In B.J. Huntley and B.H. Walker (eds), *Ecology of Tropical Savannas*. Berlin: Springer-Verlag, pp. 273–91.

Dargie, T. and Furley, P. (1994) Monitoring change in land use and the environment. In P.A. Furley (ed.), *The Forest Frontier. Settlement and Change in Brazilian Roraima*. London: Routledge, pp. 68–85.

Eden, M.J. (1974) Paleoclimatic influences and the development of savanna in southern Venezuela. *Journal of Biogeography*, **1**, 95–109.

Eden, M.J. (1986) Monitoring indigenous shifting cultivation in forest areas of southwest Guyana using aerial photography and Landsat. In M.J. Eden and J.T. Parry (eds), *Remote Sensing and Tropical Land Management*. Chichester: John Wiley and Sons, pp. 255–77.

Eden, M.J. (1990) *Ecology and Land Management in Amazonia*. London: Belhaven Press.

Eden, M.J. (1994) Environment, politics and Amazonian deforestation. *Land Use Policy*, **11**, 55–66.

Eden, M.J. and McGregor, D.F.M. (1992) Dynamics of the forest-savanna boundary in the Rio Branco–Rupununi region of northern Amazonia. In P.A. Furley, J. Proctor and J.A. Ratter (eds), *Nature and Dynamics of Forest-Savanna Boundaries*. London: Chapman and Hall, pp. 77–89.

Eden, M.J., McGregor, D.F.M. and Vieira, N.A.Q. (1990) Pasture development on cleared forest land in northern Amazonia. *Geographical Journal*, **156**, 283–96.

Egler, E.G. (1961) A zona Bragantina do Estado do Pará. *Revista Brasileira de Geografia*, **23**, 527–55.

Endler, J.A. (1982) Pleistocene forest refuges: fact or fancy? In G.T. Prance (ed.), *Biological Diversification in the Tropics*. New York: Columbia University Press, pp. 641–57.

Fearnside, P.M. (1979) Cattle yield prediction for the Transamazon highway of Brazil. *Interciencia*, **4**, 220–5.

Foresta, R.A. (1991) *Amazon Conservation in the Age of Development*. Gainesville: University of Florida Press.

Friedman, I. (1977) The Amazon basin, another Sahel? *Science*, **197**, 7.

Furley, P.A. and Ratter, J.A. (1990) Pedological and botanical variations across the forest-savanna transition on Maraca Island. *Geographical Journal*, **156**, 251–66.

Gentry, A.H. and Lopez-Parodi, J. (1980) Deforestation and increased flooding of the upper Amazon. *Science*, **210**, 1354–6.

Gómez-Pompa, A., Vázquez-Yanes, C. and Guevara, S. (1972) The tropical rain forest: a non-renewable resource. *Science*, **177**, 762–5.

Goodland, R.J.A. and Irwin, H.S. (1975) *Amazon Jungle: Green Hell to Red Desert?* Amsterdam: Elsevier.

Haffer, J. (1969) Speciation in Amazonian forest birds. *Science*, **165**, 131–7.

Hamilton, L.S. (1991) Tropical forests: identifying and clarifying issues. *Unasylva*, **42**, (166) 19–27.

Heywood, V.H. and Stuart, S.N. (1992) Species extinctions in tropical forests. In T.C. Whitmore and J.A. Sayer (eds), *Tropical Deforestation and Species Extinction*. London: Chapman and Hall, pp. 91–117.

Hills, T.L. (1964) *Progress Report, Savanna Research Project*. Unpublished report. Geography Department, McGill University, Montreal.

Huber, O., Steyermark, J.A., Prance, G.T. and Ales, C. (1984) The vegetation of the Sierra Parima, Venezuela-Brazil: some results of recent exploration. *Brittonia*, **36**, 104–39.

INPE (1992) Deforestation in Brazilian Amazonia. *Revista SELPER*, Santiago, **8**, 35–6.

Kubitzki, K. (1985) The dispersal of forest plants. In G.T. Prance and T.E. Lovejoy (eds), *Amazonia*. Oxford: Pergamon Press, pp. 192–206.

Lean, J. and Warrilow, D.A. (1989) Simulation of the regional climatic impact of Amazon deforestation. *Nature*, **342**, 411–13.

Lovelock, J.E. and Kump, L.R. (1994) Failure of climate regulation in a geophysiological model. *Nature*, **369**, 732–4.

Lucas, R.M., Honzak, M., Foody, G.M., Curran, P.J. and Corves, C. (1993) Characterizing tropical secondary forests using multi-temporal Landsat sensor imagery. *International Journal of Remote Sensing*, **14**, 3061–7.

Mausel, P., Wu, Y., Li, Y., Moran, E.F. and Brondizio, E.S. (1993) Spectral identification of successional stages following deforestation in the Amazon. *Geocarto International*, **4**, 61–71.

Medina, E. and Silva, J.F. (1990) Savannas of northern South America: a steady state regulated by water-fire interactions on a background of low nutrient availability. *Journal of Biogeography*, **17**, 403–13.

Milliken, W. and Ratter, J.A. (1989) *The Vegetation of the Ilha de Maraca. First Report of the Vegetation Survey of the Maraca Rainforest Project (INPA/RGS/SEMA)*. Edinburgh: Royal Botanic Garden.

Molion, L.C.B. (1976) *A Climatonomic Study of the Energy and Moisture Fluxes of the Amazonas Basin with Considerations of Deforestation Effects*. São José dos Campos: Instituto de Pesquisas Espaciais.

Myers, N. (1979) *The Sinking Ark. A New Look at the Problem of Disappearing Species*. Oxford: Pergamon Press.

Myers, N. (1988) Tropical forests and their species. Going, going ... ? In E.O. Wilson (ed.), *Biodiversity*. Washington DC: National Academy Press, pp. 28–35.

Myers, N. (1991) Tropical forests: present status and future outlook. *Climatic Change*, **19**, 3–32.

Nepstad, D.C., Uhl, C. and Serrão, E.A.S. (1991) Recuperation of a degraded Amazonian landscape: forest recovery and agricultural restoration. *Ambio*, **20**, 248–55.

Nobre, C.A., Sellers, P.J. and Shukla, J. (1991) Amazonian deforestation and regional climate change. *Journal of Climate*, **4**, 957–88.

Prance, G.T. (1973) Phytogeographic support for the theory of Pleistocene forest refuges in the Amazon basin, based on evidence from distribution patterns in Caryocaraceae, Chrysobalanaceae, Dichapetalaceae and Lecythidaceae. *Acta Amazonica*, **3**, 5–28.

Prance, G.T. (1990) Consensus for conservation. *Nature*, **345**, 384.

Rawitscher, F. (1948) The water economy of the vegetation of the 'Campos Cerrados' in southern Brazil. *Journal of Ecology*, **36**, 237–68.

Richey, J.E., Nobre, C. and Deser, C. (1989) Amazon river discharge and climate variability: 1903 to 1985. *Science*, **246**, 101–3.

Salati, E., Marques, J. and Molion, L.C.B. (1978) Origem e distribuição das chuvas na Amazônia. *Interciencia*, **3**, 200–5.

Salati, E., Vose, P.B. and Lovejoy, T.E. (1986) Amazon rainfall, potential effects of deforestation, and plans for future research. In G.T. Prance (ed.), *Tropical Rainforests and the World Atmosphere*. Boulder: Westview Press, pp. 61–74.

Scott, G.A.J. (1978) Grassland development in the Gran Pajonal of eastern Peru. A study of soil-vegetation nutrient systems. *Hawaii Monographs in Geography*, **1**, 1–187.

Shukla, J., Nobre, C. and Sellers, P. (1990) Amazonian deforestation and climate change. *Science*, **247**, 1322–5.

Sioli, H. (1973) Recent human activities in the Brazilian Amazon region and their ecological effects. In B.J. Meggers, E.S. Ayensu and W.D. Duckworth (eds), *Tropical Forest Ecosystems in Africa and South America: A Comparative Review*. Washington DC: Smithsonian Institution, pp. 321–34.

Smith, N.J.H. (1992) Despite grave losses of forest, some good things are happening in Amazonia. *Environmental Conservation*, **19**, 294–5.

Sommer, A. (1976) Attempt at an assessment of the world's tropical moist forests. *Unasylva*, **28**, 112–13, 5–24.

Sternberg, H. O'R. (1987) Aggravation of floods in the Amazon river as a consequence of deforestation? *Geografiska Annaler*, **69A**, 201–19.

Turner, J.R.G. (1971) Studies of Mullerian mimicry and its evolution in Burnet moths and heliconid butterflies. In R. Creed (ed.), *Ecological Genetics and Evolution*. Oxford: Blackwell Scientific Publications, pp. 224–60.

Uhl, C. (1982) Recovery following disturbances of different intensities in the Amazon rain forest of Venezuela. *Interciencia*, **7**, 19–24.

Uhl, C. (1988) Restoration of degraded lands in the Amazon basin. In E.O. Wilson, (ed.), *Biodiversity*. Washington DC: National Academy Press, pp. 326–32.

Uhl, C., Buschbacher, R. and Serrão, E.A.S. (1988) Abandoned pastures in eastern Amazonia. 1. Patterns of plant succession. *Journal of Ecology*, **76**, 663–81.

Uhl, C. and Vieira, I.C.G. (1989) Ecological impacts of selected logging in the Brazilian Amazon: a case study from the Paragominas region of the State of Pará. *Biotropica*, **21**, 98–106.

Wilson, E.O. (1988) The current state of biological diversity. In E.O. Wilson (ed.), *Biodiversity*. Washington DC: National Academy Press, pp. 3–18.

5

Degradation of Tropical Rain Forest in Southeast Asia: Taxonomy and Appraisal

Alan Grainger

Large areas of the world are covered not by primary ecosystems but by vegetation types that have been degraded by human impact. From a cultural biogeographical perspective, degraded ecosystems are likely to be the rule rather than the exception in the modern world, but they have received little academic study except as stages in the succession process. Consequently, biogeography has no coherent framework available for mapping the actual distribution of world vegetation, which consists of a complex mosaic of primary and degraded ecosystems, rather than merely a collection of climax vegetation types.

Concern about vegetation degradation in the tropics has grown rapidly in recent years. In the humid tropics, concern about the fate of the tropical rain forest has switched from a preoccupation with deforestation to a consideration of the impacts that leave behind degraded forest. In dry areas, desertification has been portrayed as the sum of soil degradation and vegetation degradation (Grainger, 1990), though, because of the neglect of vegetation degradation, assessments of the extent of soil degradation are more advanced, despite the difficulty of monitoring over large areas.

Mapping the distribution of degraded vegetation has now become a priority for those studying the erosion of biodiversity and the role of tropical forests in global climate change. This chapter proposes an initial set of degradation indicators and a taxonomy of degraded forest types as a contribution to this endeavour. The social aspects of degradation, which are just as important as its physical characteristics, will be examined and a case study from South and Southeast Asia will be presented to show the possibilities for appraising degraded forests on a large scale and the merits of an integrated, socio-physical approach to evaluating the role of such forests in global environmental change.

Deforestation and Degradation

The most drastic human impact on tropical rain forests is deforestation, defined here as 'the temporary or permanent clearance of forest for agriculture or other purposes' (Grainger, 1986, 1993). However, selective logging, the predominant forest management practice in these areas, does not clear the forest and cause deforestation, but merely removes some commercial trees.

Another term is needed to represent the effects of logging and other impacts which merely modify forest cover, whether or not clearance is involved. The term now widely used for this purpose is vegetation degradation. It is defined here as a 'temporary or permanent reduction in the density, structure, species composition or productivity of vegetation cover' (Grainger, 1992, 1993). Deforestation is an extreme case of degradation, as it temporarily reduces the density of the vegetation cover to zero. After deforestation the vegetation cover may increase as a result of planting or natural regeneration, although often the result is inferior, or degraded, relative to what was there before.

Degradation was defined earlier by Serna (1986) as

> the reduction in, or impairment of, the capability of a forest area to produce wood due to the influence of outside factors, particularly human activities. Degradation involves no diminution of area but only a decrease in the productivity of the resource.

However, this definition was intended to serve the needs of forestry applications and is more limited in scope than the one above.

There are three main causes of degradation: first, extraction, e.g. selective logging, which may be followed by either tree planting or managed natural regeneration; second, forest damage resulting from natural or indirect human causes, like drought or pollution; and third, clearance followed by managed regeneration, planting or interrupted succession.

Two Perspectives on Degradation

Degraded forest and vegetation have been traditionally portrayed from a successional perspective as secondary forest in a state of transition between the two extremes of cleared land and climax forest (Bazzaz and Pickett, 1980; Ewel, 1980; Denslow, 1987). Logged forest and shifting cultivation forest fallow are both types of secondary forest. The scrubby or weedy wastelands that characterize areas under encroaching cultivation or highly intensive shifting cultivation are plagio-climax (or fire-climax) vegetation.

Such a 'bottom-up' approach provides a very inadequate description of the various types of degraded forest, since it focuses on comparing the state of vegetation with its poor condition preceding succession, rather than the ideal climax cover. It also emphasizes natural regeneration processes rather than the human processes that cause degradation. Repeated human intervention is acknowledged as impeding succession, but is represented as more of a static constraint than a complex set of human interactions with biotic processes. The literature on secondary forests also focuses on natural successional forest rather than on vegetation that has been wholly or partially planted artificially. For these reasons and since it is now vital to monitor the loss of biodiversity or biomass due to human impact, an alternative 'top-down' approach to degradation is desirable (Brown and Lugo, 1990) and is adopted here.

Major Physical Degradation Indicators

The four key characteristics of degradation in the above definition – structure, density, species composition and productivity – provide a good starting point for defining a set of physical indicators that can be used to distinguish between different types of degraded forest.

Vertical Structure

The vertical structure of degraded forest often differs from that of primary forest in that some of the multiple layers of vegetation are lacking. For example, logging removes much of the emergent tree layer, and forest fallow has a low canopy.

Density

Density has five main characteristics: canopy closure, canopy quality, fragmentation, tree density and biomass density. Not all of these indicators are applicable to all degraded forest types.

1. Canopy closure
Though undisturbed forest contains a good sprinkling of gaps due to natural causes, e.g. tree mortality, lightning damage, fires, etc., young degraded forest generally exhibits more gaps in the canopy. For example, logging

removes commercial trees and produces significant gaps, though they will be filled as soon as younger trees grow tall enough.

2. Canopy quality
The quality of leaf cover may be abnormal in degraded forest. Drought, fire, insect pests, pollution, defoliants and other natural and human hazards can kill or damage foliage, and even change the surface coverage of epiphytes.

3. Tree density
Degradation may also reduce the density of trees. In an extreme case, overlogging might convert a closed forest into an open forest, in which tree crowns are widely dispersed over a lower layer of grassland or shrubs, as in savanna woodlands. When forest is replaced by forest fallow, on the other hand, the site is quickly colonized, giving a high overall stem density, even though most of the trees have a small diameter and low basal area. High canopy closure and leaf area index are achieved within a short time, but the canopy height remains quite modest.

The density of large trees is also important, because owing to their larger diameter and higher wood density they normally contain a disproportionate share of total biomass (Weaver and Murphy, 1990; Gillespie *et al.*, 1992). For example, in well-stocked forests on flat, rolling or mountainous terrain in Sarawak, large trees (above 70 cm diameter breast height) accounted for 38 to 48 per cent of the total biomass of 350 to 400 Mg ha^{-1} (Brown *et al.*, 1994).

4. Biomass density
Biomass density, the mass of vegetation (or carbon) per unit area, is reduced as a forest is degraded, as the overall density of trees declines, as large trees are removed or as mature forest is replaced by a bushy forest fallow of smaller trees. Monitoring changes in biomass density is crucial to achieving a proper understanding of the contribution of tropical forests to the greenhouse effect.

5. Fragmentation
Changes in canopy closure are also apparent in a wider spatial context. Human impact may have fragmented the former homogeneous spatial distribution of undisturbed forest to give a complex mosaic of undisturbed forest, degraded forest, gaps and clearances. This is listed below as a type of degraded forest cover in its own right. Even if substantial patches of undisturbed forest remain, forest cover in the entire area is clearly degraded. The

more fragmented the forest, the more likely it is that the remaining forest will be disturbed eventually (Brown *et al.*, 1994).

Species Composition

Degradation can lead to a temporary or longer-term change in species composition compared with that of undisturbed forest. This may occur in various ways, for example, the extraction of some primary species, the growth of secondary species, or the planting of monocultures or species mixtures. This has obvious consequences for the forest's biological diversity, and will also change its productivity indicators; for example, the net biomass accumulation in a young secondary forest is higher than in a primary forest.

Animal species composition should be considered too, since the commercial poaching of wildlife degrades forest just as much as overlogging. Removing animals directly or by modifying their habitats, as in logging, will affect the reproductive systems of trees that depend on the animals for pollination or seed dispersal, and so will ultimately degrade the ecosystem as a whole.

A Taxonomy of Degraded Forest Types

If the distribution of degraded forest is to be mapped, it is important to devise a taxonomy that is comprehensive and functional. In this section a taxonomy is proposed that has been developed for degraded forest in Southeast Asia. Six categories of degraded forest are distinguished using the indicators described above (Table 5.1).

1. Extractive forest
Extractive forest has experienced the removal of trees or herbaceous plants. It differs from other types of degraded forest in that degradation has not involved prior clearance.

1a. Logged forest Tropical rain forest that has been selectively logged experiences temporary degradation. This is reversed when the forest regenerates. The canopy usually remains, though many emergents may have been removed (Grainger, 1993). Gaps left in the canopy after tree removal are normally filled by juvenile trees. The canopy is also likely to have been disturbed erratically by falling trees and other trees brought down in the logging process. Sometimes overlogging damages the canopy so much that

Table 5.1 A taxonomy of degraded forest types.

---

1 Extractive forest
   (a) Logged forest
   (b) Other extractive forest

2 Damaged forest

3 Regenerated forest
   (a) Long-rotation shifting cultivation regrowth
   (b) Short-rotation shifting cultivation forest fallow
   (c) Managed forest

4 Planted vegetation
   (a) Enriched forests
   (b) Agroforestry combinations
   (c) Tree-crop plantations

5 Interrupted successional vegetation

6 Dispersed forest

---

the result is more like an open than a closed forest. In both cases, clearing forest for skidding trails and roads may create an extended pattern of striations.

The degrading impacts of logging are apparent in the hill forests of Peninsular Malaysia, where logging has reduced the mean tree density from 250 to 289 stems ha^{-1} to only 220 and the mean biomass density from 279 to 191 Mg ha^{-1}. The effect of illegal logging was evident in another officially 'undisturbed' primary hill forest, where the mean biomass density declined from 330 to 279 Mg ha^{-1} between 1972 and 1982 (Brown *et al.*, 1994).

A key aim of selective logging is to leave a sufficient number of medium- and small-sized trees of commercial species behind to provide the basis for the next harvest. However, careless logging and overlogging may mean that the number of trees remaining is less than desirable. The continuing extraction of an elite group of commercial species over successive harvest cycles can result in the populations of these species becoming dominated by individuals with inferior form and wood quality, thereby reducing the forest's ability to generate good quality timber. Eventually, the populations could be totally depleted, thereby lowering overall biodiversity.

1b. Forest modified by other extractive forest uses   The other principal extractive forest use, the collection of minor forest products, normally has a minimal impact on the ecosystem, though sometimes whole trees are felled to exploit fruits or orchids in their canopies, and tapping wild trees for latex and resins can affect their productivity as timber trees. In other situations, latex-producers and other desirable trees are planted to increase their frequency in the forest. Another kind of extraction results from livestock grazing and browsing. This has long been a source of degradation in the dry tropics, but is now having a serious impact in parts of the humid tropics, notably the Philippines.

In tropical rain forests, fuelwood is usually collected as dead wood rather than cut from live trees, so there is minimal degrading impact. It can become more serious, however, when forest resources decline to a low level. Illegal cutting of fuelwood in government forest reserves can also have a detrimental impact on timber production.

2. Damaged forest

This is a miscellaneous category of forest degradation resulting from impacts of various kinds and durations. Natural hazards, like drought and wild fires, cause foliar damage and tree mortality, as can insect pests and air and water pollution. Sometimes the effects are widespread, but on other occasions they are quite localized. Plantations are more susceptible to insect attacks than natural forests, though both are vulnerable to drought. Over 3 million ha of tropical rain forest were damaged by drought and fire in eastern Borneo in 1982–83 (Lennertz and Panzer, 1983). Mangrove forests close to mines and to oil and natural gas wells have been degraded by heavy metal and tar pollution in rivers. Tropical forests do not yet suffer from air pollution to the same extent as temperate forests, but this will surely change as tropical countries industrialize.

3. Regenerated forest

3a. Long-rotation shifting cultivation regrowth The third main type of degraded forest results from regrowth after clearance. In traditional shifting cultivation, regrowth follows the path of succession towards climax forest. It seldom reaches the climax, but after twenty to thirty years tree height approaches that of mature forest.

3b. Short-rotation shifting cultivation forest fallow The vertical structure of forest fallow in short-rotation shifting cultivation is distinct from that of high forest. There is usually just a single layer, the canopy, and it is much lower than in high forest. The horizontal structure of a short-rotation shifting cultivation area is interesting. The small annual clearances of 1 to 2 ha lead to a chequered pattern of clearances and forest fallow stands, each with a different height, depending on when the site was last abandoned by farmers. This form of degradation is relatively permanent, lasting at least as long as shifting cultivation continues in the area. Even if a piece of land is finally abandoned, reduced fertility and the growth of weeds can prevent forest regenerating in the foreseeable future.

The tree density in young secondary forest is much lower than in primary forest. For example, in Sri Lanka, young secondary forest had only 192 stems ha^{-1} compared with 384 to 450 stems ha^{-1} in primary forest. However, in older secondary forest, this had risen to 386 stems ha^{-1} (Brown *et al.*, 1989). In Peninsular Malaysia, the mean biomass density for forest fallows was 145 Mg ha^{-1}, only half that of primary forest (Brown *et al.*, 1994).

The species composition of forest fallow is dominated by pioneer species and is less diverse than mature rain forest. If cultivators plant fruit trees before abandoning the plot, as a sign of occupation, or nitrogen-fixing trees, to speed up fertility regeneration, there is a marked increase in biodiversity (Grainger, 1993). In general, however, species diversity is markedly lower in forest fallow. Even in ten-year-old forest fallow in Venezuela, for example, species density was only a third of that in mature forest; after twenty years, this had risen to 60 per cent, and after sixty years it was still only 73 per cent. Regaining the original species density can take eighty years or more (Saldarriaga *et al.*, 1986).

Low species diversity reflects the small number of pioneer species, such as *Macaranga* spp., that invade the area. They eventually give way to primary forest species as these colonize the patch and the canopy closes. This is often a slow process and depends on the species composition of the surrounding vegetation. Fruit trees and other planted species will be present in degraded forests from the early stages. On the other hand, shifting cultivation may increase the biodiversity of a large area, since the pioneer species present in regenerating patches will supplement the primary species in undisturbed areas.

3c. Managed forest   In some countries, foresters use silvicultural methods to intervene in the succession process to favour regrowth of commercial timber species and to maintain forests at that stage of succession in which growth rates and economic rates of return are high. Managed forests are therefore artificially constrained secondary forests.

4. Planted vegetation
4a. Enriched forests   Another group of degraded vegetation types results from intentional tree planting. The simplest form of planting is the enrichment of natural forests, for example by planting commercial tree species as part of a silvicultural system, or planting fruit or rubber trees in the forest for subsistence purposes.

4b. Agroforestry combinations   Alternatively, various tree crop mixtures may be planted, either on cleared land or in a forest. These mixtures, which may or may not include forest trees and under-planted pastures, come under the general heading of agroforestry systems, the collective name for practices that combine growing trees with field crops and/or raising livestock. The canopy is usually undulating and there are multiple vegetation layers. These plantations vary greatly in size, but are usually much smaller than mono-culture tree-crop plantations.

4c. Tree-crop plantations   Plantations of tree crops like oil palm or timber species are widespread in the humid tropics. Many of them are mono-cultures, and so have a highly degraded species composition. They are characterized by a uniform canopy, and the lack of a grass layer and a second layer of trees. They often cover substantial areas. The biomass density of a non-timber tree crop plantation is lower than that of a forest, typically 50 to 70 per cent of a natural primary forest (Dale *et al.*, 1993). Degradation is permanent.

5. Interrupted successional vegetation
This group is structurally distinct from the others, as it has a low tree component and is dominated by grassland. For example, in encroaching cultivation, the third main type of shifting cultivation and the most destruct-ive (Eckholm, 1976; Grainger, 1993), land is cleared of forest and then farmed for more than two years and only abandoned when fertility is exhausted and weed growth is widespread. Alternatively, forest regeneration is continually interrupted by burning, as in the African savannas. The result in both cases is a more extreme form of degradation than that caused by other shifting cultivation practices. The vegetation cover is a weedy shrubland or grassland rather than a degraded forest. It consists of large areas of weedy grasses, such as *Imperata cylindrica*, punctuated by dispersed trees. Biomass density may be less than 10 per cent of that in primary forest (Dale *et al.*, 1993).

6. Dispersed forest
The final group of degraded vegetation types comprises landscapes with mixtures of forest trees, tree crops and weedy grasslands in a dispersed, heterogeneous mosaic. It can include mosaics of encroaching cultivated land, forest fallow and forest. The overall tree cover is generally higher than in interrupted successional vegetation; degradation is extensive but not as extreme, and land use practices are more sustainable.

Dispersed forest is an advanced case of forest fragmentation. As people encroach upon the outer boundaries of a block of forest, the ratio between the total perimeter and area increases. With greater clearance along the edges comes the risk of illegal logging in the remaining forest, reducing its biomass density. For example, in Peninsular Malaysia between 1972 and 1982, the mean perimeter/area ratio of unlogged primary hill forest increased from 1.55 to 1.92, and the mean biomass density declined from 330 to 279 Mg ha^{-1} (Brown *et al.*, 1994).

The Appraisal of Degraded Forests

Extent of Degradation

Current estimates of the extent of degraded forest, and of degraded lands generally, are still very inaccurate. Degraded forest accounted for a third of all closed woody cover in tropical Asia in 1980, with 60 million ha of logged forest and 69 million ha of forest fallows compared with a total of 375 million ha of closed forests and fallows (Lanly, 1981). The only available estimate of the degradation rate in Asia is that provided by the logging rate for tropical moist forest, namely, 1.3 million ha yr^{-1} in the 1970s, rising to 1.8 million ha yr^{-1} in the 1980s (Grainger, 1986; FAO, 1993).

Remote Sensing Techniques

Lack of data reflects the fact that monitoring forest degradation with remote sensing techniques is more difficult than monitoring deforestation. The visible and infra-red sensors carried on satellites are more sensitive to leaf area than to ecosystem structure or biomass density. Non-degraded forest is therefore easily confused with tree crop plantations and small areas of forest fallow, since their canopies have similar leaf areas. However, large blocks of forest fallow can be distinguished, and interrupted successional vegetation and dispersed forest with low tree cover are easier to identify. One reason for including agricultural land uses in the above taxonomy is the need to cope with such complexity in practical monitoring situations.

Structural differences, like those between logged and unlogged forest, are only apparent when using higher resolution systems, for example aerial photography or radar sensors. The growing number of radar satellites should soon make routine monitoring of forest degradation possible. Degradation can also be monitored indirectly, for example by monitoring the perimeter/area ratio of remaining forest and sampling biomass density (Brown *et al.*, 1994).

Estimating the Degree of Physical Degradation

In an attempt to improve existing estimates and appraisal techniques, Iverson *et al.* (1993) estimated the extent of degraded forests in continental South and Southeast Asia, their degree of degradation, and their suitability for being rehabilitated so as to mitigate global climate change by increasing biomass density and sequestering surplus carbon from the atmosphere.

The first phase of the study involved mapping the distribution of degraded forest that was *technically suitable* for carbon enhancement because its biomass density was sub-optimal (Iverson *et al.*, 1993). The degree of degradation was assumed to be proportional to the difference between potential and actual biomass density. Potential biomass density in existing forests was represented by a potential carbon sequestration index (PCSI) on a scale of 1 to 100. This was estimated using a formula that included climate and other environmental variables, and was mapped using a geographic information system (GIS) to combine regional maps of these variables. Actual biomass density was represented by an actual carbon sequestration index (ACSI), estimated in two ways: (a) by a model based on climate and settlement data; and (b) by a global vegetation index derived from low-resolution satellite data. The two estimates showed good correlation. The distribution of existing forests was identified from an FAO vegetation map that had been derived mainly by visual interpretation of medium-resolution Landsat imagery.

The degree of degradation was estimated by calculating the difference between the ACSI and the PCSI layers on the GIS. This resulted in a map showing forests grouped into six degradation classes (Figure 5.1). The biomass density of the lowest class was 50 Mg C ha^{-1} less than its potential value, the highest more than 250 Mg C ha^{-1} lower. The higher the class, the more degraded the forest, the higher its potential carbon uptake, and the higher its technical suitability for carbon enhancement.

About 27 per cent of all the forest in the study area – 45 to 50 million ha, according to alternative scenarios – was in the three highest degradation classes. This proportion was compatible with the estimate referred to above. More than 80 per cent of the highly degraded forest was in India, Burma and Peninsular Malaysia (Figure 5.1), although the apparent high degradation in Peninsular Malaysia probably has much to do with better data availability and the high level of management for timber production. Cambodia, Laos and Vietnam had the least degraded forests. Half of all forest in the three highest degradation classes was closed forest, and another third was shared between open forest and forest fallow.

Compared with a mean biomass density across the region of 185 Mg C ha^{-1}, the average for degraded forest was 88 Mg C ha^{-1}. Just under one third of all forest could sequester less than 50 Mg C ha^{-1}, but almost two-thirds could sequester 50 to 150 Mg C ha^{-1}, and so make a meaningful contribution

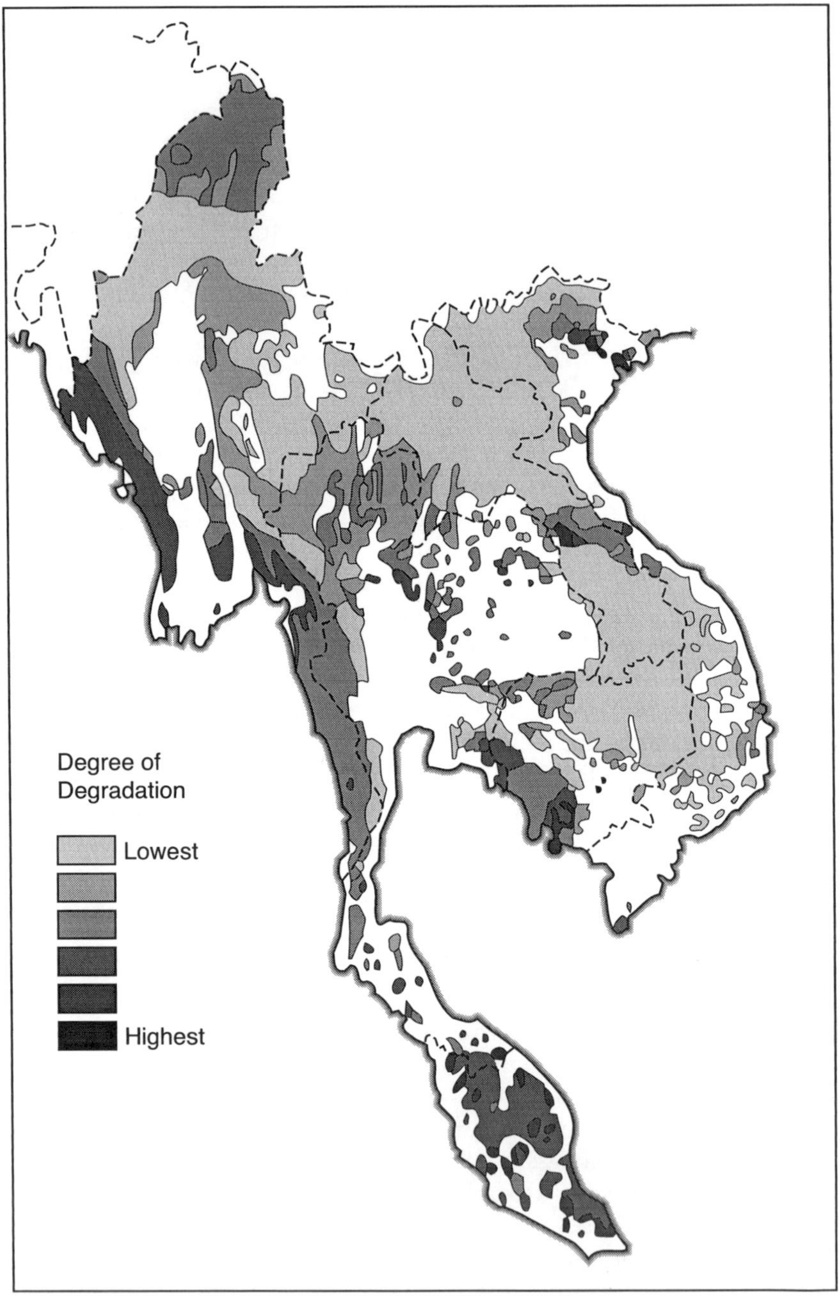

Figure 5.1 The degree of forest degradation in continental Southeast Asia. *Source:* after Iverson *et al.* (1993).

to mitigating global climate change (Iverson *et al.*, 1993). These estimates are very preliminary, since they depend on unreliable maps of forest cover and land use, and there are large errors associated with estimating the ACSI and PCSI and using regression equations to predict their regional distributions.

Including the Social Dimensions of Degradation

The main limitation of the first stage of this study was that it focused purely on physical aspects of degradation and neglected social aspects. The aim of the second phase was to estimate the proportion of technically suitable degraded forest that was actually available for rehabilitation given the social, economic and political constraints on land use change.

For degraded land to be available in this way, it must be degraded in relation to cultural as well as physical criteria. While degradation is usually thought of as a perjorative term, so far in this chapter it has been used solely to indicate inferiority relative to ideal physical norms. However, only some physically degraded lands are also culturally degraded, in that they are not economically or socially useful. Physically sub-optimal vegetation may well be culturally optimal, since some degree of physical degradation can be a necessary consequence of particular management systems. The challenge is to devise techniques which can, within the limitations imposed by poor spatial data availability, incorporate an additional social dimension which makes it possible to assess whether the land is culturally degraded and available for restoration.

In the second phase of the Asian study, the socio-economic and physical dimensions of degradation were blended in two stages to allow a comprehensive appraisal. In the first stage, degraded forest was separated into its different uses. Forest currently being managed in some way for timber production was regarded as available for improved management, and degraded land with a low tree cover was regarded as available for afforestation. But land under more intensive agricultural use, including tree-crop plantations, other cropland, permanent pasture or forest fallow, was regarded as unavailable. The second stage of the analysis refined these estimates to take a fuller account of the social, economic and political constraints on availability (Grainger, 1996).

Conclusions

This chapter has described the main types of degraded forest in Southeast Asia using a taxonomy based on a set of key degradation indicators. Degraded forests have long been neglected and there is much to be learned about them. Given the growth of research on global environmental change and the development of new remote sensing and GIS techniques, continuing advances are expected. Meanwhile, the limited data that are available will have to be used to produce the best available estimates of degraded forest distribution, using approaches similar to those outlined here.

This field offers many research challenges, one of the most important being to learn more about the socio-economic dimensions of degradation and how to superimpose these on the physical distributions of degraded forest. This is not a trivial exercise, as culturally modified forest is subject to both social and physical forces, and human-vegetation interactions are reciprocal, not uni-directional. In order to assess the potential for restoring degraded forest to mitigate global climate change, for example, both constraining and modifying pressures must be included in the group of socio-economic attributes.

Given the traditional approach of biogeography to degraded vegetation, it will take time to develop a fresh cultural biogeographic view of the subject, but initial research on global environmental change offers hope that progress can be made.

References

Bazzaz, F.A. and Pickett, S.T.A. (1980) Physiological ecology of tropical succession: a comparative review. *Annual Review of Ecology and Systematics*, **11**, 287–310.

Brown, S., Gillespie, A.J.R. and Lugo, A.E. (1989) Biomass estimation methods for tropical forests with applications to forest inventory data. *Forest Science*, **35**, 881–902.

Brown, S., Iverson, L.R. and Lugo, A.E. (1994) Land use and biomass changes in Peninsular Malaysia during 1972–1982: a GIS approach. In V. Dale (ed.), *Effects of Land Use Change on Atmospheric CO₂ Concentrations: Southeast Asia as a Case Study*. Berlin: Springer Verlag, pp. 117–43.

Brown, S. and Lugo, A.E. (1990) Tropical secondary forests. *Journal of Tropical Ecology*, **6**, 1–32.

Dale, V., Houghton, R., Grainger, A., Lugo, A. and Brown, S. (1993) Emissions of greenhouse gases from tropical deforestation and subsequent uses of land. In National Research Council, *Sustainable Agriculture and the Environment in the Humid Tropics*. Washington DC: National Academy Press, pp. 215–60.

Denslow, J.S. (1987) Tropical rainforest gaps and tree species diversity. *Annual Review of Ecology and Systematics*, **18**, 421–51.

Eckholm, E. (1976) *Losing Ground*. New York: W.W. Norton.

Ewel, J.J. (ed.) (1980) Tropical succession. *Biotropica*, **12** Supplement, 1–95.

FAO (1993) *Forest Resources Assessment 1990: Tropical Countries*. Rome: United Nations Food and Agriculture Organization.

Gillespie, A.J.R., Brown, S. and Lugo, A.E. (1992) Tropical forest biomass estimation from truncated stand tables. *Forest Ecology and Management*, **48**, 69–88.

Grainger, A. (1986) *The Future Role of the Tropical Rain Forests in the World Forest Economy*. Unpublished DPhil thesis, Department of Plant Sciences, University of Oxford.

Grainger, A. (1990) *The Threatening Desert: Controlling Desertification*. London: Earthscan Publications.

Grainger, A. (1992) Characterization and assessment of desertification processes. In G.P. Chapman (ed.), *Proceedings of Conference on Grasses of Arid and Semi-Arid Regions, Linnean Society, London, 27th February–1st March 1991*. Chichester: John Wiley, pp. 17–33.

Grainger, A. (1993) *Controlling Tropical Deforestation*. London: Earthscan Publications.

Grainger, A. (1996) Integrating the socio-economic and physical dimensions of degraded tropical lands in global climate change mitigation assessments. In M. Apps (ed.), *The Role of Forest Ecosystems and Forest Management in the Global Carbon Cycle*. Berlin: Springer Verlag, pp. 335-48.

Iverson, L.R., Brown, S., Grainger, A., Prasad, A. and Liu, D. (1993) Carbon sequestration in South/Southeast Asia: an assessment of technically suitable forest lands using geographic information systems analysis. *Climate Research*, **3**, 23–38.

Lanly, J.P. (ed.) (1981) *Tropical Forest Resources Assessment Project (GEMS)*. Vols 1–4. Rome: Food and Agriculture Organization/United Nations Environment Programme.

Lennertz, R. and Panzer, K.F. (1983) *Preliminary Assessment of the Drought and Forest Fire Damage in Kalimantan Timur*. Bonn: DFS German Forest Inventory Service.

Saldarriaga, J.G., West, D.C. and Thorp, M.L. (1986) *Forest Succession in the Upper Rio Negro of Colombia and Venezuela*. Environmental Sciences Publication No. 2694, Oak Ridge National Laboratory, Oak Ridge.

Serna, C.B. (1986) *Degradation of Forest Resources. Special Study on Forest Management, Afforestation and Utilization of Forest Resources in the Developing Regions. Asia-Pacific Region*. GCP/RAS/106/JPN, Field Document No. 15. Rome: United Nations Food and Agriculture Organization.

Weaver, P.L. and Murphy, P.G. (1990) Forest structure and productivity in Puerto Rico's Luquillo Mountains. *Biotropica*, **22**, 69–82.

6

Monitoring Sediment Yield and Logging Practices: Implications for Forest Management in Sabah, Malaysia

Tony Greer, Ian Douglas, Kawi Bidin, Waidi Sinun,
Jadda Suhaimi and Azman bin Sulaiman

Development in the form of commercial logging inevitably results in land disturbance and degradation. As forest resources diminish, more marginal and inaccessible upland regions are being exploited and cleared, as in the case of Sabah, Malaysia. Upland watersheds are fragile environments and disturbance usually causes accelerated soil erosion and sedimentation. Such disturbance will modify the design of any monitoring and management programme. Watershed research can contribute to better resource management because water and sediment discharge data from small catchments and erosion plots provide an understanding of the important erosion processes occurring in commercially logged rain forest, which, in turn, indicate how water conservation and erosion control techniques may be used by forest managers.

Aside from the geomorphological interest, the study of sediment transfer within river systems is important in applied contexts because of the effects on channel geometry and capacity and on water quality. Excess suspended material may also adversely affect fluvial ecosystems by disrupting biotic development, reducing photosynthesis, interfering with engineering projects such as irrigation schemes and hydroelectric power plants, and impairing the quality of water used for domestic and industrial purposes (Cooke and Doornkamp, 1990).

Sound land use practice can benefit from the assimilation of data from detailed research programmes. Currently, there is an expanding data base and a wide range of management options. Remedial or palliative action will depend on the willingness and ability to comply with and implement amelioration and conservation programmes, be it at a corporate or national level. Different response levels exist and these cannot be divorced from social, political and economic issues. In this chapter, procedural considerations in monitoring sediment dynamics and yield will be examined at a case-study level. It will be shown how methodological and analytical decisions can affect management practices.

Drainage Basin Monitoring and Management

Drainage basins provide an integrated system through which water and sediment are transferred. Of crucial importance to land managers is an understanding of the intimate link between cause and effect or between the different components of a system and any human interference, such as deforestation, which usually triggers a response from the system (Cooke and Doornkamp, 1990). There need to be effective linkages between erosion, sediment transport and deposition models. Precise routes of sediment through a basin are complex and the sediment budget of a system is difficult to calibrate. Consequently, at the present time, sediment-related planning and predictions of sediment movement are expressed in terms of three general measures (Cooke and Doornkamp, 1990):

1 Gross erosion – the total erosion within a basin arising from all erosion processes, expressed either by weight or in terms of ground surface lowering per unit of time.
2 Sediment yield – the total sediment outflow from a drainage basin through a particular discharge point in a specified unit of time.
3 Sediment delivery rate – the ratio between sediment yield and gross erosion within a catchment, taking into account the amount of sediment in storage.

Forest Degradation and Sediment Yield

A comprehensive review of catchment studies conducted during tropical forest conversion is presented in Bruijnzeel (1990). In discussing changes in basin sediment yield attributed to deforestation, he indicates that it is important to distinguish between surface erosion, gully erosion and mass movement, since the ability of a vegetation cover to control various forms of erosion is rather different. Bruijnzeel also stresses the importance of distinguishing between on-site erosion and off-site downstream effects. Only part of the eroded material will immediately enter the fluvial transport system, other components being detained in temporary storage sites, such as ephemeral channels and depressions, to be reworked during larger storm events. Storage opportunity generally increases as a power function of basin size.

The impacts of erosion are typically more significant on-site than downstream, where the effects may be delayed over a range of temporal scales from storm to seasonal to long-period events. According to Pearce (1986) and Hamilton (1987), there could be very little change for decades in the amounts of sediment carried by major rivers in their lower reaches, even if all anthropic erosion in the uplands were immediately eliminated, as there is so much stored sediment throughout the catchment. Thus, upland reforestation is not a panacea that will generally solve downstream problems. It is important not to raise unrealistic expectations, otherwise the credibility of watershed management and environmental planning, and possibly years of progress towards more rational land use, could be put at risk (Bruijnzeel, 1990).

Measurement of Sediment Concentration

Given the complex nature of sediment transport and the often transitional division between modes of transport, an appropriate sampling programme is critical. In the context of commercial forestry, it is not practical or necessary to carry out detailed catchment studies for all systems that are to be disturbed. Data sets may be extrapolated over regional units. The objectives of a particular programme will be biased towards the concerns of the land managers, who in turn are likely to react to government policy directives. The most immediate geomorphological impact of commercial forestry is on soil erosion, soil compaction and fluvial siltation. Thus, a monitoring programme must be designed accordingly. Instantaneous sediment discharge can be measured from the product of sediment concentration and liquid discharge. Given a continuous record of stream discharge, sediment concentration data can be integrated to obtain the sediment load over a specified period, e.g. annually. In determining sediment discharge, it is preferable to consider bedload and wash load separately since different factors govern their transport in a river (Dickenson *et al.*, 1990). There is, however, an intimate relationship between bed material, bedload and suspended load, and consequently the choice of measuring equipment will bias the fraction that is being sampled. Any device that rests on the stream bed will retain mainly bedload, and any device that protrudes upwards from the stream bed or is raised or lowered through the flow will sample mainly suspended load (Emmett, 1981).

Wash load consists of fine particles which are carried in suspension. Their discharge primarily depends on the rate of sediment supply and is a function of the factors which govern gross erosion rates within the catchment, such as rainfall, relief, soil type, vegetative cover and land use (Dickenson *et al.*, 1990). Wash load tends to be uniformly distributed through the channel section and is relatively easy to sample. The complex nature of suspended and bedload movement makes it more difficult to obtain a representative sample. Particles are not uniformly distributed with depth and a discrete sampling regime is required in order to accurately define the sediment concentration profile.

Movement of bedload and suspended load is largely governed by hydraulic parameters and so previous research has predicted sediment yield using sediment rating curve-flow duration methods. This requires the simultaneous measurement of streamflow and sediment to develop a sediment rating curve. It is then possible to extrapolate a short period of sediment sampling to much longer periods of flow data. In large rivers, where sediment concentrations usually change little over the day, these values are sometimes adequate for calculating daily suspended sediment discharge. However, in 'flashier' forest streams, there will be a tendency towards underestimating sediment load. Sediment concentrations are often poorly related to discharge due to hysteresis effects or varying amounts of wash load. As wash load is less dependent on hydraulic parameters, reliable estimates can only be made by more discrete continuous monitoring. Therefore it is preferable to measure sediment concentrations rather than attempt to estimate them from streamflow data (Dickenson *et al.*, 1990).

Forest Disturbance at Ulu Segama

Research on tropical rain forest disturbance and geomorphic processes at the Danum Valley Field Centre, Ulu Segama, Sabah began in October 1987, with the objective of establishing the effects of natural disturbance, such as gap formation and anthropic disturbance, particularly logging, on run-off and erosion. The focus of this study is the assessment of sediment sources and the measurement of suspended sediment loads in two small catchments, A and C (Figure 6.1). Catchment A, Sungai W8S5 (1.7 km^2), is in undisturbed rain forest of the conservation area, and catchment C, Sungai Steyshen Baru (0.5 km^2), is in a logging concession where extraction of timber began in January 1989.

Measurement Techniques

Stream water sampling on a per-visit basis involved collecting a 500 ml grab sample. The turbulent nature of the streams ensured that a well-mixed and representative sample was taken. Higher flow samples were collected by means of automatic liquid samplers (ALS), with 1 litre bottles, activated by a float switch during storm run-off events. After each storm, suspended sediment concentrations were obtained by filtration through 0.45 micron cellulose acetate membrane filters (Douglas, 1971).

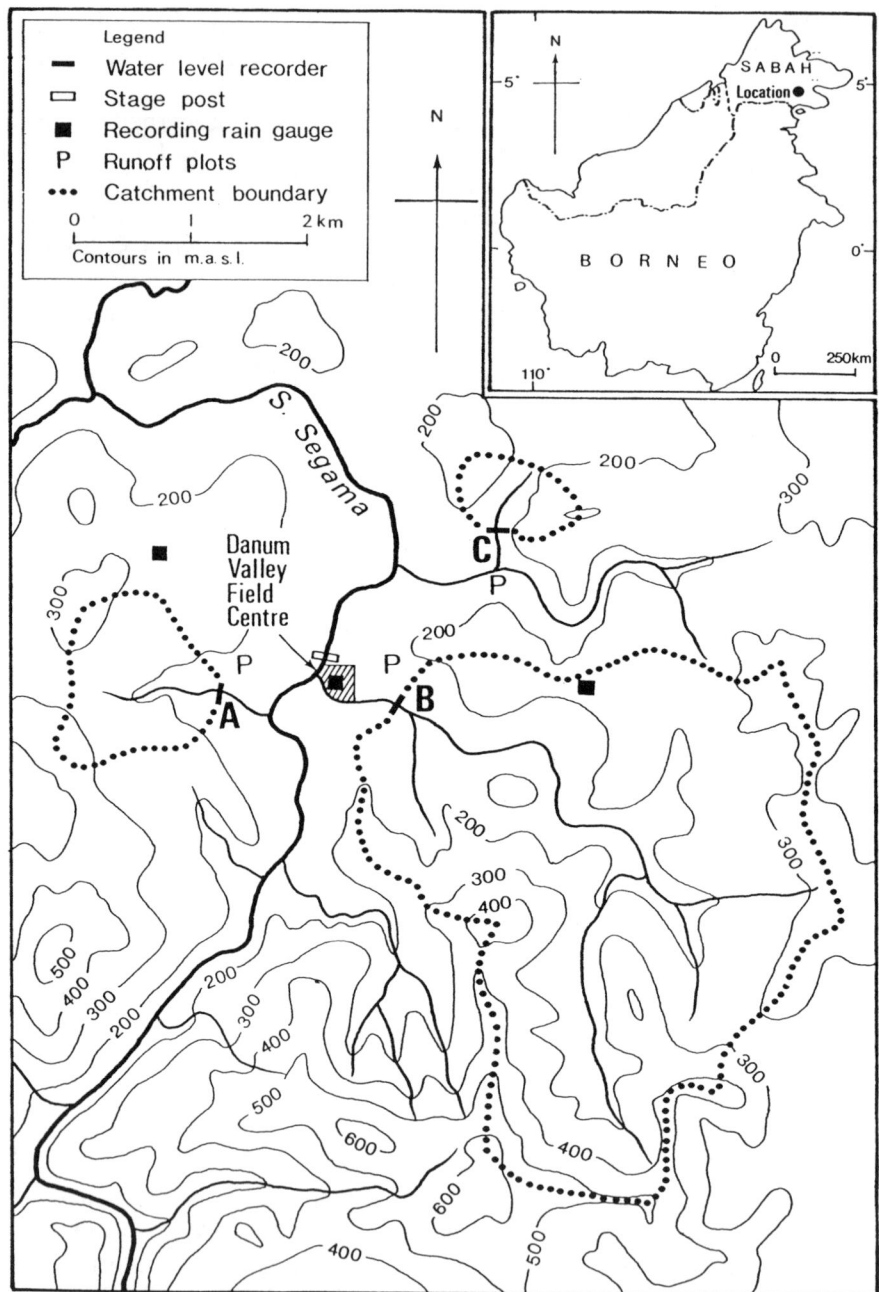

Figure 6.1 The Ulu Segama study site, Sabah. Catchment A – Sungai W8S5, Catchment C – Sungai Steyshen Baru.

Actual sediment loads were calculated from the storm sediment samples and by integrating values between baseflow samples. These loads are hereafter referred to as *measured*. At this initial stage, the sediment concentration data were grouped into three categories, namely: (a) all samples, (b) ALS samples only, and (c) grab samples only. Sediment yields were then derived by the application of the rating curve derived from the above sampling techniques to the continuous discharge record (Table 6.1).

Table 6.1 Relationships between suspended sediment concentration and discharge.

Catchment A, Sungai W8S5			
All data			
Log Conc = Log Q 0.7896	+ 2.4096	n = 1355	r^2 = 0.54
Grab samples only			
Log Conc = Log Q 0.3001	+ 1.1434	n = 213	r^2 = 0.17
Automatic sampling (ALS)			
Log Conc = Log Q 0.5614	+ 2.3815	n = 1142	r^2 = 0.26
Catchment C, Sungai Steyshen Baru			
All data			
Log Conc = Log Q 0.5656	+ 3.1884	n = 2303	r^2 = 0.49
Grab samples only			
Log Conc = Log Q 0.3311	+ 2.2863	n = 332	r^2 = 0.20
Automatic sampling (ALS)			
Log Conc = Log Q 0.1810	+ 2.9164	n = 1971	r^2 = 0.03

n = number of samples
Q = instantaneous discharge (m^3 s^{-1})
Conc = instantaneous suspended sediment discharge (mg l^{-1})

Bedload movement and solution are part of the natural denudation process and sediment yield estimates must be assessed by direct or empirical means to take these into account. The present emphasis is on suspended sediment and the effects of removing or disturbing the vegetation cover. The actual amount of bedload in streams flowing through non-montane, forested environments is small. In Java, Bruijnzeel (1983) computed bedload at 12.6 per cent of total sediment output.

It is probable that bedload increases with disturbance, as there is an increased and ready supply of a full range of particle sizes. However, there are also increased in-stream obstacles in the form of logging debris. Clear-cutting of plantation forest (*Pinus merkusii*) in Java (Bons, 1990) resulted in a two-stage increase in bedload and suspended load. The increase did not immediately follow clear-cutting, but came as delayed flushing removed the material trapped during forest clearance. As forest harvesting extends into steeper and more marginal lands, the proportion of bedload may be expected to increase.

The proportion of solute in the total load is variable. A relatively high load of 30 per cent was computed by Douglas (1973) for a tropical catchment in Queensland. For the Mondo river basin in Java, Bruijnzeel (1983) obtained results of 19 per cent. Differences in lithology, annual precipitation, vegetation and land use can affect the amount of dissolved material in the total load.

Sediment Yield

It is not the purpose of this study to evaluate the merits of sediment rating curve derivation, but to examine practical methods that can aid land management. As many recent papers have described, the sediment load of a river is likely to be underestimated by using the conventional least squares regression on the logarithm of concentration. The value given by the anti-logged regression is the geometric mean, which will necessarily be lower than the arithmetic mean. The degree of underestimation increases with the degree of scatter about the rating curve and can reach 50 per cent (Ferguson, 1986). It is sufficient to note that there are procedures to adjust for this through a correction factor (Ferguson, 1986) or by using an allometric non-linear regression model (Dickenson *et al.*, 1990). However, Walling and Webb (1988) suggest that bias-corrected rating curves do not provide accurate estimates for rivers in Devon, UK and that other sources of errors associated with rating curves are more important in producing inaccurate estimates. In the context of anthropic disturbance, it is probable that these adjustments are applicable to tropical rivers.

The measured and calculated suspended sediment loads for the two catchments for a sample month (July in 1989, 1990 and 1991) are compared in Figure 6.2. In 1989, the sediment load in catchment C (Sungai Steyshen Baru) is underestimated by the all-sample rating, ALS rating and grab rating in that order. In 1990, both the all-sample rating and ALS rating provide an overestimate compared to the measured load. The same pattern prevails in 1991. The grab rating is consistently low, reflecting its bias towards low-flow data, and hence is unacceptable except during low water periods. In catchment A (Sungai W8S5), the all-sample rating closely approximates the measured load, except for the low-flow period in 1991 when the grab sample is more representative. For catchment C, the all-sample rating and ALS rating shift from underestimate to overestimate. The same variation can be seen with catchment A.

In general, considerable discretion must be used in selecting a rating technique in computing sediment discharge data. Grab samples alone drastically underestimate catchment sediment yields and consequently gross erosion. Conversely, the ALS rating for catchment C progressively overestimates the load. This partly reflects recovery of the catchment vegetation. The recovery rating, however, can be complicated as larger storm events will still evacuate large amounts of sediment, as sources and pathways are reactivated. Vegetation recovery, especially in the pioneer

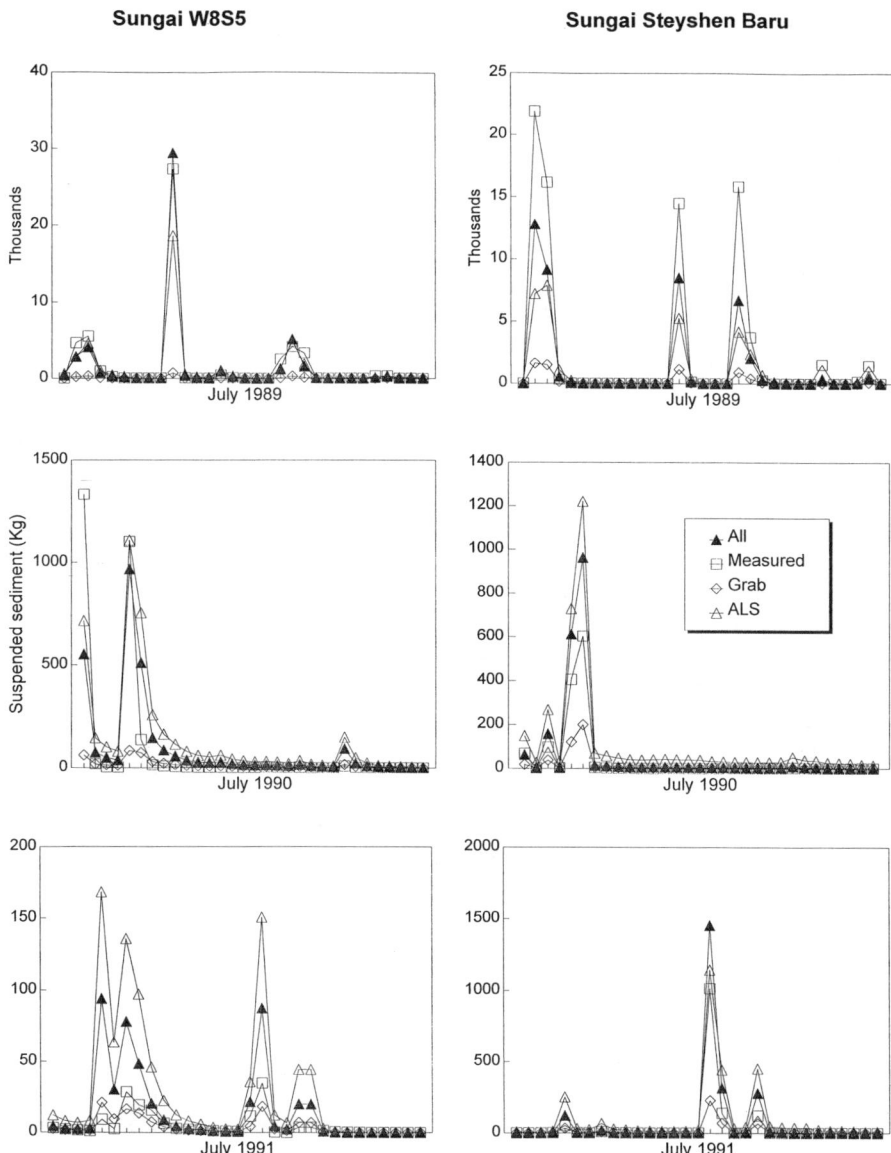

Figure 6.2 Measured and computed sediment loads for sample months (July 1989, 1990 and 1991) at Sungai W8S5 (Catchment A) and Sungai Steyshen Baru (Catchment C).

herbaceous layers, will generally start to reduce erosion rates (e.g. Thornes, 1989), but the storage of sediments introduces considerable variance. The farther downstream from the area of disturbance, the more difficult yield prediction becomes as a storage flux variable is increasingly introduced (Trimble, 1990).

Catchment condition and treatment cannot be considered on the basis of rating curve computations alone. Practice should incorporate a critical review of the rating curve on an annual division or water-year basis. When resources permit, ratings should be reviewed using a much smaller temporal unit.

Plot Studies

During the logging operation, skid trails are extended from the primary track to the cut trees. Small run-off plot experiments were set up on recently abandoned skid trails, with bounded plots to prevent run-off from sources away from the skid entering the traps. Sediment traps similar to Gerlach troughs were used, emptying into plastic collecting drums. Adjacent to the main run-off plot (plot 1), a skid trail with matching slope and soil characteristics was blocked at the top by bulldozing a sediment bar, i.e. a 2m-high earth mound (plot 2). This prevented trail and road run-off from channelling down the slope, a common phenomenon along skid trails (Swift, 1988). Plots were also established on a 1-year-old skid trail and in the nearby primary forest.

Run-off plots were not installed immediately on fresh skid trails due to continuing logging operations, but this allowed the trails to undergo some adjustment. An obvious difference after a period of only a few weeks was that vegetation, mainly in the form of grasses, began to establish on plot 2 (protected by the sediment bar). The difference has persisted so that there is now a dense impenetrable growth, whilst plot 1 remains sparsely vegetated (Anderton, 1990).

A bounded plot soon becomes a self-contained unit (Mykura, 1985), with significant effects on plant-growing conditions. Plot 1, which was unprotected, became very active with severe gullying. In July 1989, this plot yielded 190.5 t ha^{-1} of sediment compared with only 10.5 t ha^{-1} from plot 2. The latter, protected by the sediment bar, was noticeably different with rapid establishment of grasses and other herbaceous growth. Disturbed, loose soil was quickly removed in plot 1, but remained in place in plot 2 as vegetation became established. Heavy erosion continued, with most loss occurring in the first few months before regrowth became established (Sinun, 1991). The one-year abandoned track yielded 0.53 t ha^{-1}, slightly more than the 0.38 t ha^{-1} from the adjacent primary forest plot (Figure 6.3). However, by 1990, yield from the unprotected plot 1 had diminished to only 10.5 t ha^{-1}. It is important to note when using small run-off plot experiments that the stage at which the plot is established is crucial, e.g. if the plot is installed immediately after the construction of a sediment bar, it will not be possible to measure the benefits of recolonization.

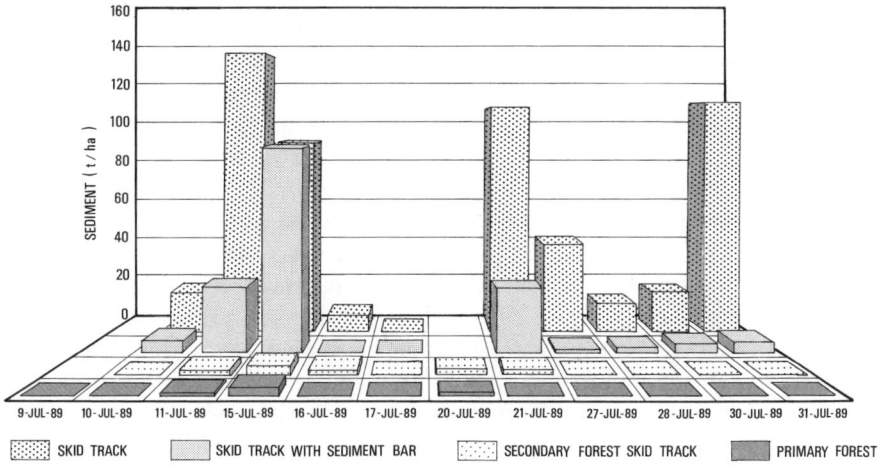

Figure 6.3 Sediment yield measured at four different control plots, Ulu Segama, Sabah.

Establishment of vegetation can be rapid in the humid tropics, provided soil surface conditions are suitable. Once grasses are established, other herbaceous growth quickly follows. The rate at which this occurs is highly site-dependent, steeper areas often supporting sparser vegetation cover than gently sloping ones. Prompt reduction of run-off, after use of roads and skid trails ceases, is a key to encouraging rapid vegetation recolonization. Earth barriers to impede run-off should be constructed as soon as possible after skidding ceases. They will be more effective in reducing erosion in the early months than tree planting, which used to be practised in Sabah.

Traffic density will affect compaction. In these experiments, skid trails were used to remove only a few logs, but major trails close to access roads experience numerous tractor movements. The type of logging operation and site characteristics mainly determine the impact of disturbance (Rice and Datzmann, 1981).

Discussion

This case study indicates how different sampling strategies and data management can produce different and possibly misleading information for resource managers. Reliable information is required for appropriate mitigative and ameliorative management to take place.

The structured removal of forest in commercial logging operations permits tighter control over environmental degradation and consequently has greater management potential than, for example, the more random effects of shifting agriculture. New logging techniques aimed at minimizing impacts on the soil are being introduced, although the institutional frameworks within which forestry is managed means that these new approaches often lack an adequate soil conservation input. Erosion studies, although recognized, are still not fully acknowledged by commercial foresters as an integral part of the resource to be managed. Accounting for soil as a natural resource and subsequently as a capital asset helps address this (Repetto *et al.*, 1989).

An understanding of sediment dynamics within a catchment is essential to sound environmental management. It is not suggested that detailed monitoring programmes are required in the day-to-day management of forested areas, but catchment research studies can provide baseline data from which regional development may benefit. Often experiments have provided inadequate sampling of storm run-off so that sediment yields are underestimated and the real magnitude of the erosion problem is overlooked. Discretionary use of appropriate monitoring and rating-curve analytical techniques gives a more representative description of catchment recovery after disturbance. Monitoring programmes are usually designed towards the user's needs. In parts of Southeast Asia where forest harvesting is taking place, population densities are low and an environmental obligation to forest and riparian dwellers is not always a priority. Monitoring criteria will also change with the stage of social and economic development. In the Segama basin, the effluent from a single oil-palm processing plant caused more concern to local people than the deforestation necessary for plantation development in the first place.

Managing tropical forest lands wisely is feasible. Soil can be conserved. Treatments that encourage water to take paths akin to those in the natural forest are available. Infiltration can be encouraged and surface run-off avoided. Protection against rain-splash erosion is possible. Much can be done in the design of logging systems and the location of access roads and skid trails, particularly by avoiding stream crossings and long runs down steep slopes. However, even where steep gradients are needed, early implementation of simple, inexpensive erosion-control techniques, such as barriers to block surface run-off, will reduce the persistent erosion that cumulatively aggravates sedimentation downstream. Such techniques are well documented (Pearce and Hamilton, 1986) and with appropriate supervision should be effective (Douglas *et al.*, 1992). Instruction of field crews and dissemination of information to forest managers are of crucial importance. Carbon-offset projects structured to subsidize the additional training and supervisory costs of reduced-impact logging operations address many of these problems and are an excellent initiative (Marsh, 1993). However, technology transfer is not always easy, whether for commercial or logistical reasons. Politics, culture and the associated problems of overseas extension workers are by no means divorced from the issues, and can prevent the widespread implementation of reduced-impact methods.

References

Anderton, S. (1990) *Logging and Soil Erosion: An Examination of the Effects of Logging on Soil Erosion in the Rain-forests of Ulu Segama, Eastern Sabah.* Unpublished BSc dissertation, University of Manchester, Manchester.

Bons, C.A. (1990) Accelerated erosion due to clear-cutting of plantation forest and subsequent 'Taungya' cultivation in upland West-Java, Indonesia. *International Association of Hydrological Sciences Publication*, **192**, 279–88.

Bruijnzeel, L.A. (1983) *Hydrological and Biogeochemical Aspects of Manmade Forests in South-central Java, Indonesia.* Unpublished PhD thesis, Free University, Amsterdam.

Bruijnzeel, L.A. (1990) *Hydrology of Moist Tropical Forests and Effects of Conversion: A State of Knowledge Review.* Amsterdam: UNESCO/IHP, Free University.

Cooke, R.U. and Doornkamp, J.C. (1990) *Geomorphology in Environmental Management: A New Introduction.* Oxford: Clarendon Press.

Dickenson, A., Amphlett, M.B. and Bolton, P. (1990) *Sediment Discharge Measurements – Magat Catchment. Summary Report: 1986–1988.* Wallingford: Hydraulics Research Report OD 122.

Douglas, I. (1971) Comments on the determination of fluvial sediment discharge. *Australian Geographical Studies*, **11**, 172–6.

Douglas, I. (1973) Rates of denudation in selected small catchments in Eastern Australia. *Occasional Papers in Geography, University of Hull*, **21**, 1–127.

Douglas, I., Spencer, T., Greer, T., Bidin, K., Sinun, W. and Meng, W.W. (1992) The impact of commercial logging on stream hydrology, chemistry and sediment loads in the Ulu Segama rain forest, Sabah. *Philosophical Transactions of Royal Society of London*, Series B, **335**, 397–406.

Emmett, W.W. (1981) Measurement of bed load in rivers. In Erosion and Sediment Transport Measurement. *International Association of Hydrological Sciences Publication*, **133**, 3–15.

Ferguson, R.I. (1986) River loads underestimated by rating curves. *Water Resources Research*, **145**, 403–17.

Hamilton, L.S. (1987) What are the impacts of deforestation in the Himalayas on the Ganges–Brahmaputra lowlands and delta? Relations between assumptions and facts. *Mountain Research and Development*, **7**, 256–63.

Marsh, C. (1993) Carbon dioxide offsets as potential funding for improved tropical forest management. *Oryx*, **27** (1), 2–3.

Mykura, H. (1985) Research notes: design and operation of a simple erosion plot. *Ilmu Alam*, **14**, 105–13.

Pearce, A.J. (1986) *Erosion and Sedimentation*. Working Paper, Environment and Policy Institute, Honolulu.

Pearce, A.J. and Hamilton, L.S. (1986) *Water and Soil Conservation Guidelines for Land Use Planning*. Honolulu: East–West Centre.

Repetto, R., Magrath, W., Wells, M., Beer, C. and Rossini, F. (1989) *Wasting Assets: Natural Resources in the National Income Accounts*. Washington DC: World Resources Institute.

Rice, R.M. and Datzmann, P.A. (1981) Erosion associated with cable and tractor logging in north-western California. *International Association of Hydrological Science Publication*, **132**, 362–74.

Sinun, W. (1991) *Hillslope Hydrology, Hydrogeomorphology and Hydrochemistry of an Equatorial Lowland Rain Forest, Danum Valley, Sabah, Malaysia*. Unpublished MSc thesis, University of Manchester, Manchester.

Swift Jr, L.W. (1988) Forest access roads: design, maintenance, and soil loss. In T. Swank and D.A. Crossley Jr (eds), *Forest Hydrology and Ecology at Coweeta*. New York: Springer-Verlag, pp. 313–24.

Thornes, J. (1989) Solutions to soil erosion. *New Scientist*, **122** (1667), 45–9.

Trimble, T.W. (1990) Geomorphic effects of vegetation cover and management: some time and space considerations in prediction of erosion and sediment yield. In J.B. Thornes (ed.), *Vegetation and Erosion*. Chichester: John Wiley, pp. 55–65.

Walling, D.E. and Webb, B.W. (1988) The reliability of rating curve estimates of suspended sediment yield: some further comments. *International Association of Hydrological Science Publication*, **174**, 337–50.

PART III

Degradation in the Drier Tropics

7

Land Degradation in Tropical Drylands

John T. Parry

Tropical drylands cover extensive areas from the hyper-arid cores of the major deserts to the marginal zones of wetter climate – the Mediterranean and monsoon woodlands and the savanna fringes of the dry sub-humid zone. Approximately 37 per cent of the world's land surface (*c.* 4850 million ha) comprise drylands. Hyper-aridity characterizes approximately 978 million ha (7.5 per cent) of this total, 1569 million ha (12 per cent) may be considered arid, and the balance is semi-arid (2305 million ha). Drylands dominate the tropical zone in Australia and Africa, with the latter exhibiting all types of land degradation in the most severe forms. It is estimated that 73 per cent of African drylands are degraded in some way, with soil erosion as the most serious problem.

Land degradation in the tropical drylands is not a new phenomenon. The basic problems are long-term, even though the ecological crisis point has been reached in the latter part of the twentieth century. As a result of the combined pressures of people and climate, the agricultural and grazing capacities of the semi-arid and dry sub-humid lands are experiencing a decline with measurable reductions in crop and fodder production (UNEP, 1984). The general consensus at the United Nations Conference on Desertification in 1977 was that, although conditions have been aggravated by drought, the problems are largely socio-economic, many of them resulting from the progressive cultural, political and economic marginalization of the people of the drylands, who are historically and geographically distant from the power and decision centres of many post-colonial territories.

It will be useful to look first at the bio-physical background which sets the constraints on human activities in the tropical drylands, and then to examine the human and economic dimension in search of prescriptions for better management and damage control.

Bio-physical Factors

A major difficulty in focusing on land degradation in the tropical drylands is the inherent variability of the ecosystems. Different climatic and biological indices have been used to measure the physical conditions, but many of these fail to capture the irregular, episodic events that dominate seasonal, annual and long-term cycles. Processes operate as non-equilibrium or multi-state systems that shift abruptly from one mode to another. The prime mover in these shifts or phase changes is the alternation between rainy and dry episodes at timescales varying from the seasonal to the long-term. Moisture inputs fluctuate widely in both space and time, and so process responses are highly dynamic at both shorter and longer timescales. Drought is a recurrent scourge of the drylands. The 'effectiveness' of rainfall, i.e. the amount available for plant growth, is dependent on the ecosystem output in the form of run-off and evapotranspiration, controlled by factors such as temperature, wind speed and vegetation cover. Vegetation removal or damage and global warming pose serious, and perhaps irreversible, threats to dryland ecosystems.

Soil is the fundamental medium for plant and crop growth. The fertility of most dryland soils is inherently low, and, although they are potentially renewable, they have such a slow rate of development (a few millimetres per century) that they should be treated as a non-renewable or fixed asset. The mechanisms for soil replacement are all too easily disrupted, and soil degradation can be impossible to halt or reverse once thresholds are crossed. Dryland soils have a low resilience and the current and future capacity of the soils to support vegetation or crops can be jeopardized by mismanagement, particularly when fungi and soil organisms are lost or the seed bank is pauperized by vegetation change. Two categories of soil degradation are identified: degradation resulting from displacement and physical loss of soil material by wind or water, and *in situ* deterioration resulting from physical and chemical processes, often induced by the pressures for greater productivity.

Soil degradation is the most crucial problem for the tropical drylands at the present day, a problem that is 'epidemic' and yet one that seems to be largely ignored. More than 1030 million ha of the world's drylands are affected by some form of soil degradation (UNEP, 1992). The African drylands have suffered the most severely, with some 320 million ha affected, or approximately 25 per cent of the total dryland area. Severe soil degradation is particularly significant in the semi-arid and dry sub-humid margins of Africa's drylands which have lately experienced an expansion of the agricultural frontier in an effort to boost food production for increased populations. The situation in Malawi is described by Soulsby (Chapter 8).

In many of the more productive dryland areas, crop yields have increased, even though improved productivity has often gone hand in hand with accelerating soil damage. Very serious soil degradation can occur without there being measurable soil loss. The symptoms of 'sick' soils are seen in declining fertility. In many areas, there are clear indicators that productivity has peaked and that nutrient and moisture availability are declining. The 'plateau' for cropping improvement has been reached and the damage resulting from compaction and burning, loss of organic matter, shrinkage, nutrient depletion, sealing, crusting and pan formation, salinization and sodication, and other toxic stresses has triggered rapid feedback mechanisms that make it virtually impossible to rehabilitate or reclaim the soils (Brown *et al.*, 1984; Carter, 1986; Nortcliff, 1986).

Human Influences

The term human-induced land degradation is a useful catch-all that brackets together the linked activities that have created many of the problems of 'sick' soils identified above. It is useful to review these activities in order to understand the linkages and to obtain an insight into the socio-economic factors that drive the processes.

Fire can originate from natural causes, such as lightning, but the vast majority of fires are deliberately set. In traditional dryland societies, fire was used in the management of savannas. There are many areas of derived, fire-climax savanna in Africa which have achieved a more or less stable state (Hopkins, 1965). Repeated, light burnings do not adversely affect soil biota or earthworms. There is no decrease in N_2 levels or organic matter, nor, in general, any reduction in soil fertility. Grass or bushland are burned for a variety of reasons. Fire is the prelude to clearing in traditional bush-fallow cultivation. It is commonly used to control bush encroachment. In well-managed pastoral societies, the loss of grazing for up to three months after firing was well worth the risk in view of the improved forage that followed. Burning was only undertaken when excess forage was available. Most burning was on a three- to six-year cycle, with more frequent firing limited to denser woodland that was being opened (Pratt, 1967). Problems arise when the recurrence interval between burns is reduced, and the ability of vegetation to regenerate from seed or root sprouting is jeopardized.

Much the same problem for vegetation sustainability is created by the demands for fuelwood. In tropical drylands, fuelwood is the major source of energy for cooking and for warmth in highland areas. Few tropical crops are palatable or free of toxic substances unless cooked, and so there is a steady demand for fuelwood. Tree and bush savanna can supply this demand when collection is limited to the domestic needs of traditional dryland societies.

Serious changes occur when there is escalating commercial demand for industrial use (charcoal) and even for export (Leach and Mearns, 1989). The fuel collection area, or fuelshed, for a small community of 250 to 300 people is approximately 50 km². However, 30 000 km² are required to meet the needs of an urban centre of 100 000 people. Nairobi and Khartoum have fuelshed radii of 250 to 300 km, the total area depleted of trees and dung covering some 200 000 km². Such massive demand has serious carry-over effects: woodland regeneration is impossible, soil fertility is threatened, and soil loss due to wind and water is greatly increased.

Overgrazing can result in many of the same problems as clear-cutting for fuelwood. Traditional nomadic pastoralism seldom results in overgrazing because the animals are moved in sympathy with local and seasonal conditions. In much the same way that wildlife responds to forage gradients and episodic variations in regional productivity in an open system, pastoral peoples in traditional systems are generally well-tuned to their environment and less ecologically damaging than many sedentary societies. The problems arise when too many animals are being grazed as a hedge against mortality in dry periods or when the grazing areas are restricted by conflicts in land use or government policy. Decrease in vegetation cover is a direct result of overgrazing. In addition, there are secondary effects, such as soil compaction, and the encroachment of noxious and unpalatable grasses and shrubs that find niches in degraded ecosystems. Some secondary effects of overgrazing have serious long-term implications. An increase in dust levels could modify regional climate by producing permanent inversion layers, thereby limiting air mass exchanges and reducing precipitation (Bryson, 1967).

Deforestation – the removal of the natural vegetation cover – is unfortunately a concomitant of many government schemes to bring savanna woodland and scrubland under cultivation. Clearance is achieved rapidly and permanently using powerful mechanized equipment which can strip and grub the vegetation and then break up grass-root mats. Particularly in the African drylands, extensive changes in land use are in progress driven by population pressure, the degradation of existing agricultural lands and sheer desperation. People and governments have no other options. The savanna areas are under siege as sedentary farming practices are imposed on what have been pastoral areas. This is exemplified by Darkoh (Chapter 10) in the drylands of Kenya. The conflict is often sharpened by the fact that the soils with the greatest potential for cultivation support some of the better grasslands, often reserved by pastoralists to provide grazing in dry periods. There are serious long-term implications for all the people involved in these changes. As the amount of land available for traditional pastoralism is reduced by encroachment, the checks and balances of grazing systems that prevent land degradation are removed. In the absence of alternate grazing areas, good management is impossible and overgrazing of the available areas is inevitable (Young and Solbrig, 1993).

Running parallel to many of the land degradation processes reviewed above is population pressure. Yet no simple relationships exist between increased population, reduced productivity and land degradation. As Barrow (1991) argues, the challenge is to establish what the 'critical population' of both humans and animals is for each ecosystem. In the African Sahel, the number of people has increased at the rate of 2.5 per cent *per annum* since 1970. Some parts of the Sahel are severely stressed, whereas others are coping with the increases. It is necessary to look behind the generalities and find the specific linkages. It is increasingly realized that people must be considered integral components in dryland ecosystems, and the interactions between local communities, local resources and the broader national policies of 'development' must be subject to careful scrutiny.

There are some obvious causes of land degradation originating in political conflicts, which, although not easily remedied, are at least acknowledged as destructive. Civil wars and land disputes involving the forcible removal of people and the destruction of property and resources have devastated many parts of Africa from Biafra to Somalia, Namibia to Ruanda. Less media coverage has been given to deliberate state policies of forcing pastoral peoples to abandon their traditional ways of life and their lands, and to conform to the norms of an agrarian society. Yet the roots of these conflicts are the same – the attempt to stamp out cultural identity that is at variance with that of the ruling power. Ideological prejudice is a powerful force; it is hoped the need for survival will be even more persuasive.

It is all too easy for bureaucrats to ignore local needs and to assume that increased productivity and economic development come from government rather than the people themselves. Yet it is clear that the ecological constraints and social value systems are different in dryland areas, and so new cross-discipline paradigms that will promote interactions between local communities, the natural resource base, and the socio-political situation in the country should be developed at the local level and not superimposed from above. People are an integral part of all dryland systems and it is clear that successful management of these areas requires careful fostering of traditional values and a recognition of local diversity.

One characteristic feature of many degraded tropical drylands is a lack of formally defined land use and grazing rights. The problems in Zimbabwe are reviewed by Elliott (Chapter 9). The loose arrangements that constitute tenure rely upon common interests, effective use of the available resources, and husbanding rather than exploitative use of the ecosystem. As Young and Solbrig (1993) point out, one of the important steps in the development process in dryland areas is the formal documentation and registration of traditional land rights. This is an effective way of protecting communities and maintaining a way of life that is in tune with the sensitivity of dryland ecosystems.

Following directly from land rights is the need for the greater devolution of power and therefore management control to regional and local government. Often local communities have a far better appreciation of the problems and issues involved in good land management. The conflicts and contradictions in management policies imposed by central government are exemplified in Zinyama's analysis (Chapter 12) of the situation in Zimbabwe. With a holistic approach, there is far less likelihood of inappropriate and damaging policy decisions. Another aspect of management is equally important. Drylands are characterized by diversity of terrestrial conditions, climatic regimes, cultural and historical traditions, and economic patterns. Diversity presents problems in the planning process since it demands flexible management strategies that respond to both regional differences and time-dependent variables, such as the availability of water and grazing.

Conclusion

Many of the economic and ecological problems of the world's tropical drylands have common roots. However, they are still poorly understood and, because the areas are often marginal to the national territory and the main thrust of the economy, they have been ignored or grossly exploited because of unrealistic expectations with regard to their economic potential. It is difficult and ill-advised to provide a formula for management in dryland areas, but there are some general recommendations that have broad applicability:

1   The different faces of land degradation need to be recognized: loss of economic potential, loss of biodiversity, and, most serious of all, breakdown in ecological linkages and function.
2   A systems approach to natural processes and local problems seems to offer the best hope for understanding both ecosystems and dryland production systems.
3   In the absence of global or even regional solutions to dryland management problems, it is essential to build an understanding using local pilot projects that are based on specific ecosystem situations, community expertise and community co-operation.
4   National and international land monitoring schemes are urgently required to identify and keep track of various types of land degradation. A careful and critical assessment of remote sensing capabilities is necessary to identify optimum sensor systems and time-lapse intervals. The pioneering work of the National Remote Sensing Agency of India is described by Nagaraja and Gautam (Chapter 11).

5   Land management must be capable of accommodating local complexity and variability. Planning strategies should incorporate multi-spatial and temporal dimensions which are not constrained by national or ecological boundaries.

6   The failure of many conventional programmes emphasizes the need to involve local communities directly, providing opportunities and incentives to promote spontaneous responses. Both local and international non-governmental organizations have a major role to play in opening up new development options for the peoples of dryland areas.

References

Barrow, C.J. (1991) *Land Degradation: Development and Breakdown of Terrestrial Environments.* Cambridge: Cambridge University Press.

Brown, L.R., Chandler, W., Flavin, C., Postel, S., Starke, L. and Wolfe, E. (1984) *State of the World: A Worldwatch Institute Report on Progress towards a Sustainable Society.* New York: W.W. Norton.

Bryson, R.A. (1967) Possibilities of major climatic modification and their implications. *Bulletin of American Meteorological Society,* **48**, 136–42.

Carter, L.W. (1986) *Environmental Impacts of Agricultural Production Activities.* Chelsea: Lewis Publishing.

Hopkins, B. (1965) *Forest and Savanna.* London: Heinemann.

Leach, G. and Mearns, R. (1989) *Beyond the Woodfuel Crisis: People, Land and Trees in Africa.* London: Earthscan Publications.

Nortcliff, S. (1986) Soil loss estimation. *Progress in Physical Geography,* **10**, 249–55.

Pratt, D.J. (1967) A note on the overgrazing of burned grassland by wildlife. *East African Wildlife Journal,* **5**, 178–9.

UNEP (1984) *General Assessment of Progress in the Implementation of the Plan of Action to Combat Desertification 1978–1984.* Nairobi: United Nations Environment Programme.

UNEP (1992) *World Atlas of Desertification.* London: United Nations Environment Programme/Edward Arnold.

Young, M.D. and Solbrig, O.T. (eds) (1993) *The World's Savannas: Economic Driving Forces, Ecological Constraints and Policy Options for Sustainable Land Use.* Carnforth: Parthenon Publishing.

8

An Assessment of Vegetation Cover and Soil Erosion Hazard in Malawi Using Landsat MSS Imagery

John A. Soulsby

Soil erosion has long been recognized as one of rural Africa's most serious problems, but lack of trained personnel and the limited technical and financial resources available have meant that systematic studies to understand soil erosion have rarely been conducted. In particular, one of the major influences on soil erosion, the soil's vegetation cover, has been extremely difficult to assess. This study of rural Malawi offers a technique for evaluating vegetation cover using satellite imagery and shows how it might be incorporated into an overall scheme to monitor soil erosion.

Background to Soil Erosion in Malawi

Malawi's population in 1990 was 8.75 million with an annual growth rate of 3.52 per cent (World Resources Institute, 1992). Of this population, 92 per cent live in rural areas, but the distribution is uneven, with greater totals and growth rates prevailing in central and southern Malawi. According to FAO (1981), the annual increase of land under cultivation is about 3.5 per cent, most of which is taking place at the expense of land under woody vegetation that is neither protected by local custom nor official designation. Some 50 per cent of Malawi's land area is covered by indigenous woodland, and, of this, 20 per cent is protected as forest reserves with an additional 20 per cent being designated as National Parks or game reserves. The cleared woodland normally passes into permanent cultivation; shifting cultivation is very localized today, but, as Badcock (1949) has shown, the degraded nature of many woodlands is a legacy from past shifting cultivation, a practice which led to the designation of forest reserves as early as 1924.

Fuelwood accounts for almost 90 per cent of Malawi's primary energy supply (O'Keefe and Munslow, 1984), and the large-scale deforestation involved removes the protective cover from soils that are often highly erodable and increases erosional run-off on steep slopes. The tobacco curing industry of the central region uses 52 per cent of Malawi's total fuelwood requirement, demand continues to rise, and by the end of the century the potential fuelwood supply will have been halved (O'Keefe and Munslow, 1984). Forest and game reserves are under threat through illicit felling, and further forest degradation occurs as a result of fire. Burning is part of the savanna management tradition which not only destroys woodland during the severe fires that occur in September and October, but also exposes large areas of bare soil which are then remarkably vulnerable to erosion at the start of the rainy season in November.

A final trigger for the onset of soil erosion is the overstocking of pasture, as cattle numbers increased at an annual rate of about 5 per cent in the early 1980s. The main cattle pastures are on *dambos* that flood in the wet season but provide dry season pasture, whilst dryland grazings provide a pasture supply during the wet season. Concentrations of stock around waterholes cause excessive trampling and eventual *dambo* erosion. There is no voluntary control on cattle numbers; this is seen as contrary to traditional concepts of communal grazing rights.

Soil Erosion and Conservation Awareness in Malawi

Detailed studies of soil erosion and conservation in Malawi are few in number. Badcock (1949) has commented on the various agents of soil erosion, and, in two published reports on the physical environment of northern and central Malawi, scattered comments are made on the need for conservation measures on various soil groups (Young and Brown, 1961; Brown and Young, 1965).

Cumulative soil loss and run-off on maize plots have been monitored in central Malawi by Weil (1982), and, as Figure 8.1 shows, December is the month of maximum rainfall energy and soil loss. Weil's work also shows that soil loss increases markedly when maize plots are weeded during this period. With the onset of the rainy season in mid-November, unweeded plots will develop a substantial vegetation cover by mid-December, and at this point soil loss will have virtually ceased. Thus, Weil (1982) reports a run-off from weeded plots of 17 cm compared with 11 cm from unweeded plots, resulting in a soil loss of 12.1 t ha^{-1} and 4.5 t ha^{-1} respectively.

Awareness of the need for soil conservation is variable. On the major integrated rural land development schemes, such as that at Lilongwe, soil conservation measures are included at the planning stage (Douglas, 1988).

Large-scale farming on estates accounts for about 3 per cent of the cultivated area of Malawi, and here soil conservation techniques are practised as the norm. The principal concern thus lies with the smallholder section of the rural economy. This sector is characterized by holdings of about 1.5 ha in size, which are often fragmented into three plots per holding (National Statistical Office, 1970). This pattern of fragmentation, with its haphazard

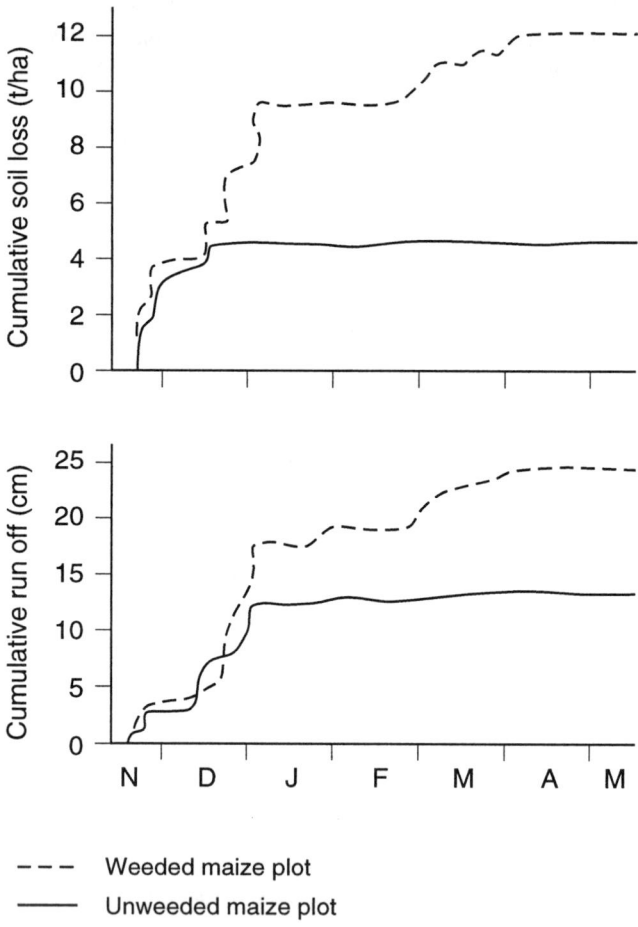

- - - Weeded maize plot

—— Unweeded maize plot

Figure 8.1 Cumulative soil loss by erosion during the 1978–79 growing season, Bunda, Malawi. *Source*: after Weil (1982).

boundaries, produces a serious erosion potential as it is difficult to organize co-ordinated soil conservation measures in such landholding circumstances. Population pressure often leads to a monoculture of maize, and crop rotation is often haphazard or non-existent.

Approaches to the Problem

As noted above, there has been little in the way of systematic investigation to discover the spatial extent of soil erosion in Malawi or to produce a technique for monitoring future erosion. This study aims to develop a methodology and techniques for estimating soil erosion hazard.

Four main factors affect soil erosion, namely, rainfall erosivity, soil erodability, slope and vegetation cover. These have been integrated into a Universal Soil Loss Equation to give the following relationship (Mitchell and Bubenzer, 1980):

$$A = 0.224 \, R \, K \, L \, S \, C \, P$$

where A = total soil loss
R = rainfall erosivity
K = soil erodability
L = slope length
S = slope gradient
C = cropping management factor
P = erosion control practice factor

This model has been used extensively to predict soil loss in the USA, but its operational effectiveness is limited in developing countries, as data for many of the variables are not available. Researchers in Africa (Stocking and Elwell, 1973; Roose, 1975) have found that an intensity-based rainfall erosivity index is significantly correlated with mean annual rainfall totals. Using values from 27 rainfall recording stations in central and southern Malawi, an index of erosivity and rainfall aggressiveness was obtained by implementing Fournier's (1960) empirical relationship between mean annual sediment yield, rainfall, altitude and slope:

$$\text{Log } Qs = 2.65 \log \frac{p^2}{p} + 0.46 \, (\log H) \, (\tan S) - 1.56$$

where Qs = mean annual sediment yield (g m^{-2})
p = highest mean monthly precipitation (mm)
P = mean annual precipitation (mm)

H = mean altitude (m)
S = mean slope (degrees)

Rainfall erosivity values are given for the test location in the lower Shire valley (Figure 8.2).

Soil erodability in the Universal Soil Loss Equation is a complex variable dependent upon texture, structural stability, shear strength, infiltration capacity, and organic and chemical content. Data for these variables are only available in generalized form in Brown and Young (1965), but an impression of the relative erodability of various soil groups is obtainable from descriptive accounts. In general, latosols, ferruginous soils, and weathered, leached and ferralitic soils have low erodability. Lateritic soils have moderate erodability, whilst alluvial calcimorphic soils, mopanosols, lithosols and regosols have high erodability. Vertisols and hydromorphic clays containing swelling montmorillonitic clay are inherently unstable and have a severe erodability potential. Maps of soil groups have been obtained from the National Atlas of Malawi (Green and Soulsby, 1983), and details for the area in the lower Shire valley are shown in Figure 8.2.

No data on slope lengths are available for substitution in the Universal Soil Loss Equation, but slope values were obtained from the Natural Regions and Areas maps (DOS, 1965). The slopes were regrouped into four gradient intervals, and their distribution for the lower Shire valley area is shown in Figure 8.2.

The Universal Soil Loss Equation also incorporates a cropping factor based on a variety of parameters, including rotations, tillage practices, and land cover characteristics at various growth stages. These data are not available for Malawi, but Elwell and Stocking (1976) have investigated the role of vegetation in the erosion process in Zimbabwe and propose percentage vegetal cover as an alternative to the cropping management factor. As Figure 8.1 indicates, the development of a vegetation cover after the onset of the wet season offers protection against soil erosion. Elwell and Stocking (1976) have demonstrated that, when vegetation cover falls to less than 30 per cent, rapid increases in run-off and soil loss occur. As cropping management factors and erosion control management factors are not available for many developing countries, including Malawi, some indication of the protection afforded by a vegetation cover would seem to be a valid alternative to the C and P factors in the Universal Soil Loss Equation. There is little conventional land-cover information for Malawi, and, since land cover probably has the greatest range of effects on erosion, a regular monitoring and assessment procedure for this variable is needed. An approximation of the degree of protection afforded to the soil by land cover was sought using Landsat MSS imagery to enable a land-cover assessment to be input into the overall erosion hazard model.

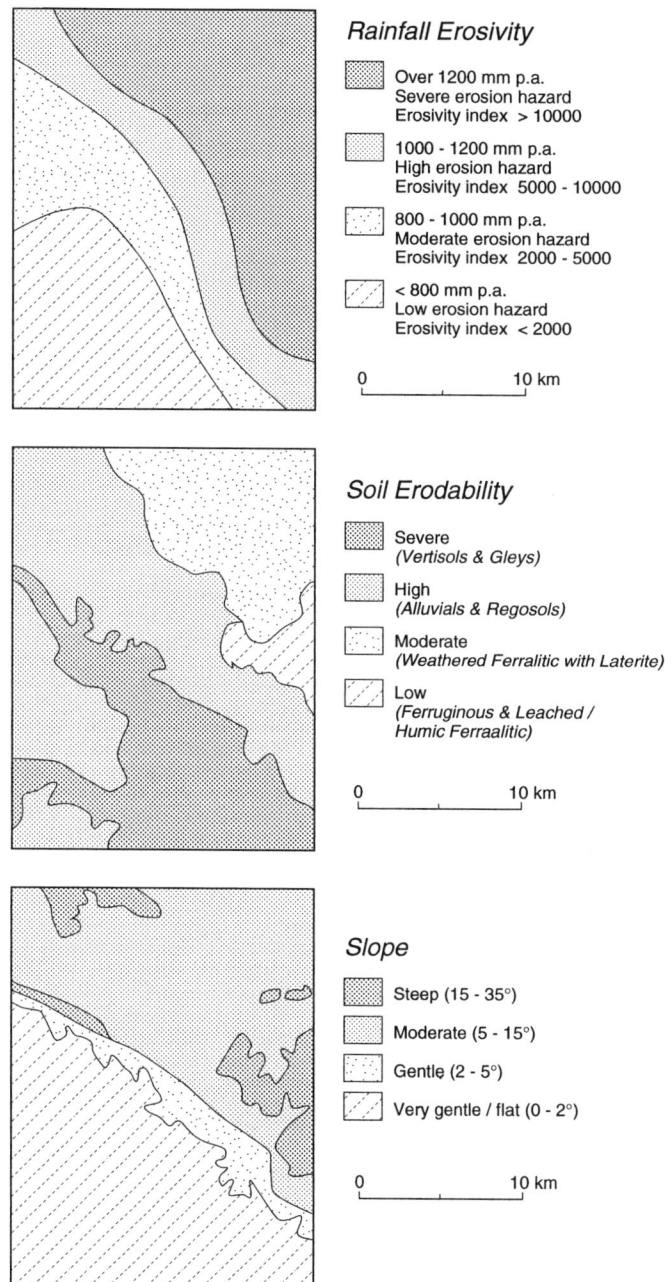

Rainfall Erosivity

Over 1200 mm p.a.
Severe erosion hazard
Erosivity index > 10000

1000 - 1200 mm p.a.
High erosion hazard
Erosivity index 5000 - 10000

800 - 1000 mm p.a.
Moderate erosion hazard
Erosivity index 2000 - 5000

< 800 mm p.a.
Low erosion hazard
Erosivity index < 2000

0   10 km

Soil Erodability

Severe
(Vertisols & Gleys)

High
(Alluvials & Regosols)

Moderate
(Weathered Ferralitic with Laterite)

Low
(Ferruginous & Leached /
Humic Ferraalitic)

0   10 km

Slope

Steep (15 - 35°)

Moderate (5 - 15°)

Gentle (2 - 5°)

Very gentle / flat (0 - 2°)

0   10 km

Figure 8.2 Rainfall erosivity, soil erodability, and slope values for the lower Shire valley, Malawi.

Image Analysis Procedures

Landsat MSS imagery (mainly cloud-free) was obtained for dates covering an eight-year period at different seasons. False colour composites using bands 4 (green), 5 (red) and 7 (infra-red) were produced and enhanced by a manual contrast stretch. Visual inspection of the imagery does not allow the identification of soil erosion *per se* owing to the limited extent of erosion scars in comparison with the Landsat MSS pixel size. Furthermore, the preponderance of bare cultivated soil on the December imagery meant that unam-

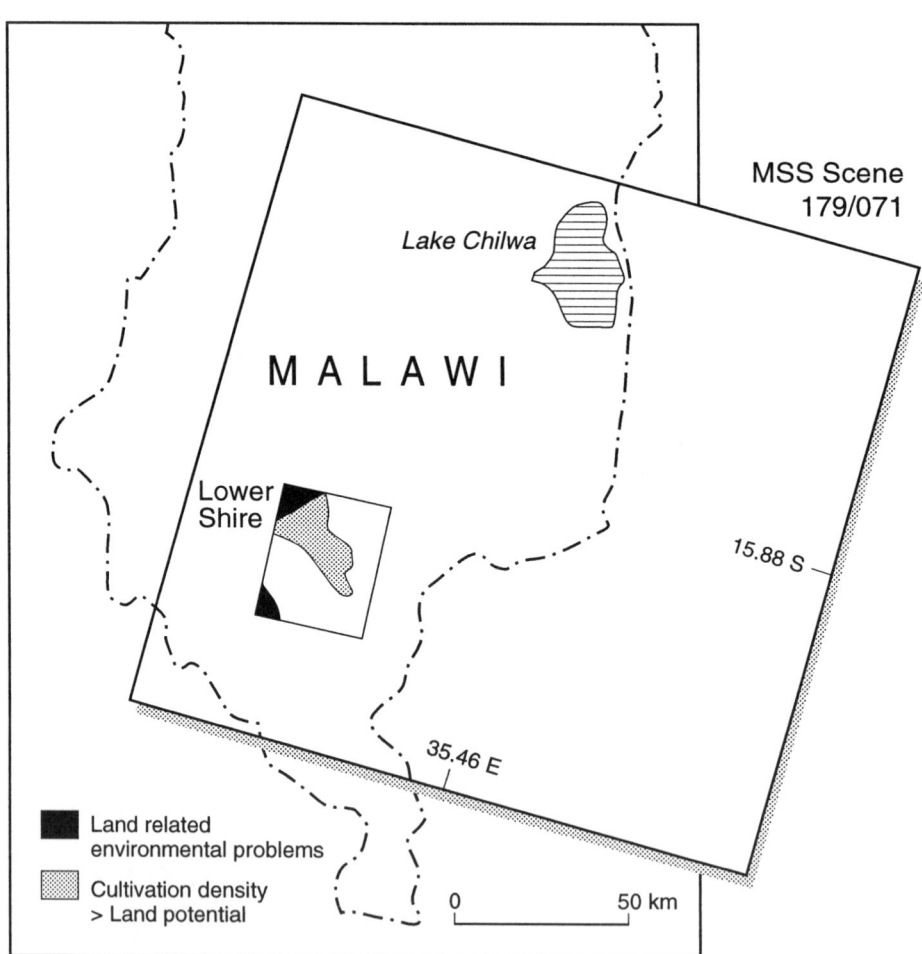

Figure 8.3 Map of southern Malawi showing location of lower Shire valley extract from Landsat MSS image.

biguous evidence of soil erosion was hard to find. The aim was therefore to develop a technique which was easy to implement using a minimum amount of image processing and which made the best use of local knowledge and documentary sources.

Ground truth was provided by the 1:50 000 national topographic map series (published 1970–74), which identifies twelve vegetation categories. This was supplemented by the 1:1 million Biotic Communities map in the National Atlas of Malawi, which shows sixteen major categories of vegetation. Most of these communities have been greatly modified by human activities and pressures: the Cultivation Density map in the National Atlas records large areas of central and southern Malawi with cultivation densities in excess of 60 per cent.

As it is not possible to investigate every Landsat scene at full resolution, samples were taken from areas which appeared to be at risk from soil erosion, having regard to the previously determined indices of rainfall erosivity, soil erodability, slope and population pressures. A further selection was made which encompassed the majority of land cover types. Combinations of the Agricultural Potential and Cultivation Density maps in the National Atlas were used to highlight areas where cultivation density exceeded agricultural potential and where environmental pressures were specifically land-related (Soulsby, 1992). This resulted in twelve extracts being selected. The present study reports on the erosion-monitoring technique as it was applied to one extract from the lower Shire valley in southern Malawi (Figure 8.3).

Approaches to Land Cover Mapping

The aim of the study is to separate different degrees of protection afforded by land covers to the soil and, as a corollary, represent different degrees of erosion hazard. Four qualitative classes of vegetation cover were established, namely (a) high, (b) moderate, (c) low, and (d) very low, in order to allow combination with the rainfall erosivity, soil erodability and slope maps.

Anderson *et al.* (1976) have proposed a resource-oriented land cover classification for use with remotely sensed data. The value of an hierarchical form of classification is stressed where possibilities exist to combine classes, extend the classification over large areas, compare imagery at different seasons, and allow comparisons to be made with future data sets. On the basis of these guidelines and with an attempt to maintain compatibility with ground information, a classification was devised and land cover types were assigned to one of the four erosion hazard ratings indicated (Table 8.1).

Spectral response curves, using a range of between 90 and 700 sample points, were then constructed for both June and December imagery to ensure

Table 8.1 Classification of land covers for use with Landsat imagery and inferred erosion hazard ratings.

Major division	Subdivision	Inferred erosion hazard rating
Forest/woodland	montane evergreen	very low
	closed-canopy forest	very low
	open-canopy savanna	low
	plantations	very low
Rangeland	montane grassland	low
	tree/scrub grassland	moderate
Agricultural land	irrigated crops	very low
	tea	very low
	non-irrigated mixed cultivation	very low
	dead standing vegetation	moderate
Open/bare land	burnt vegetation	moderate
	bare soil	high
Wetland	seasonally wet marsh grass	moderate
	perennially wet marsh grass	low

spectral separability of cover types. The range of spectral values, or digital numbers (DNs), was some 28 per cent smaller in the June (dry season) imagery, owing to the lower sun angle which produces greater shadowing effect and a less vigorous and less complete vegetation canopy with lower infra-red reflectance. The higher dust content of the atmosphere also contributes to lower reflectance values at this season. Nevertheless, the additional data from the June imagery confirmed the general accuracy of the initial December classification of land cover types.

The radiance values of bands 5 and 7 provide the greatest vegetation cover type discrimination, and a two-dimensional feature space plot for the December imagery was produced (Figure 8.4). The soil and vegetation lines are clearly shown, but the broad band on the soil axis may represent the development of a weed cover or a very low-cover/low-vigour vegetation (Gratz and Gentle, 1982). From the feature space plot, spectral separation of

Table 8.2 Separability of selected cover type pairs.

Cover type		Landsat band % separability		
		4	5	6
1 Wet burnt vegetation	D	n/s	n/s	n/s
and bare wet soil	J	n/s	n/s	n/s
2 Plantation woodland and	D	n/s	n/s	n/s
open-canopy woodland	J	66	66	66
3 Plantation woodland	D	100	100	100
and bare wet soil	J	100	100	n/s

D = December   J = June   n/s = not separable

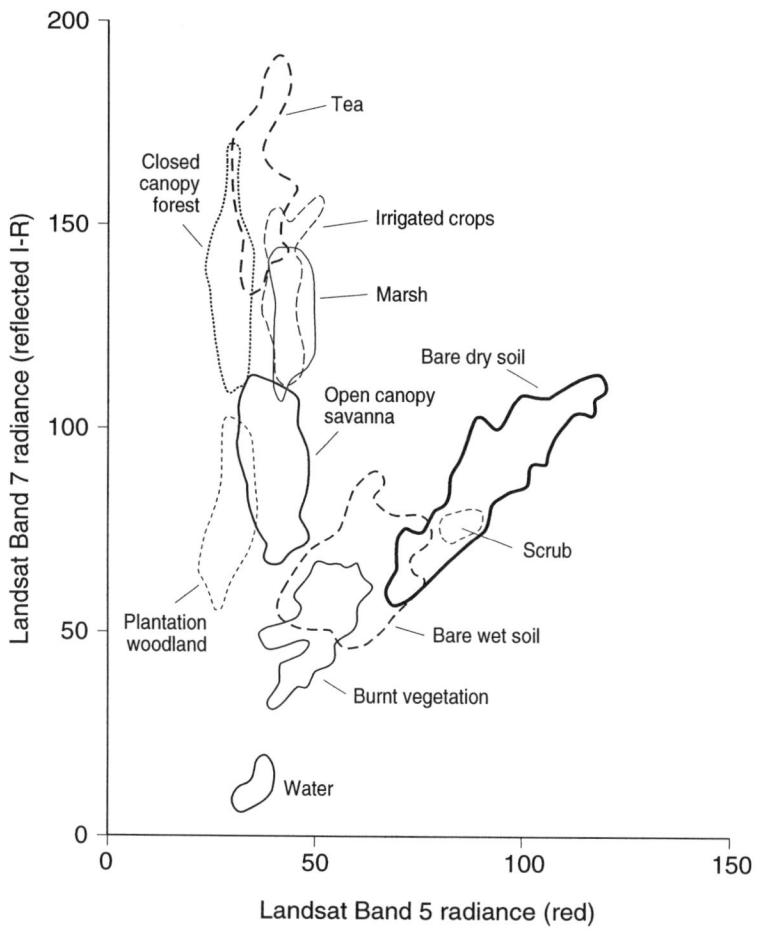

Figure 8.4 Feature space plot for Landsat MSS bands 5 and 7 (22 December, 1981), lower Shire valley, Malawi.

certain cover types is difficult as overlap occurs between irrigated sugar and marsh, but this is not necessarily detrimental as most overlapping cover types were assigned to the same erosion hazard ratings. Overlapping cover types were aggregated into nine land cover assemblages, and spectral signatures for these derived from known data sets. The range, mean, and standard deviation of these data were then used in an attempt to derive separability. Examples are given in Table 8.2: in example 1, the cover types cannot be separated on any band of the imagery at any date, and the problem is compounded by the two cover types having different erosion hazard ratings. Other cover types, however, are separable by either season or band.

Mapping Vegetation Cover and Erosion Hazard from Landsat Imagery

Two different techniques were used to examine these mapping problems:

1 Box (parallelepiped) classification. Landsat MSS data can be classified into known cover types using a box classification (Lillesand and Kiefer, 1987). Each cover type is then assigned an erosion hazard rating, using essentially a qualitative assessment of vegetation cover. The success of this approach depends upon the separability of cover types that present different degrees of erosion hazard, the accuracy of the cover type classification, and the correct interpretation of erosion hazard presented by the land cover type recognized. The feature space plot shown in Figure 8.4 was used to derive limits for the box classification.

2 Vegetation indexing. Indexing reduces MSS data to a single number, obtained by ratioing bands 7 (infra-red) and 5 (red). The index strongly correlates with percentage ground cover through biomass, plant height and leaf area index (Tucker, 1979). Band 7 is preferred to band 6, as the latter experiences more atmospheric scattering at the shorter infra-red wavelength. The ratioed values can be represented as a straight line from the origin in the feature space plot. A clear indication of the cover type values and their relationship to the box classification can thus be obtained. Cover assemblages are then grouped into erosion hazard classes as shown in Table 8.3 and Figure 8.5.

This method depends upon the ability of the vegetation index to represent vegetation cover and of the interpreter to threshold the vegetation groups in a meaningful way. As Figure 8.5 shows, the main difficulty is distinguishing between high and moderate erosion hazards. High-hazard, bare dry soil was separated from moderate-hazard, bare wet dark soil by superimposing the mapped areas of high and of moderate erosion hazard rating. The boundary was then drawn where the band 5 (red) radiance for bare dry soil exceeded

Table 8.3 Land cover assemblages.

Land cover assemblage	Examples of land cover type	Erosion hazard
High-cover/high-vigour	closed canopy forest plantation woodland tea irrigated sugar marsh	very low
Medium-cover/ medium-vigour	open-canopy savanna montane grassland	low
Low-cover/low-vigour	bare dry soil bare wet soil burnt vegetation	high/moderate

the digital number (DN) value of 60 (on a scale of 0 to 255). This value appears to separate the bare moist red soils of high erosion hazard from the bare wet dark soils and wet burnt vegetation that rate a moderate erosion hazard.

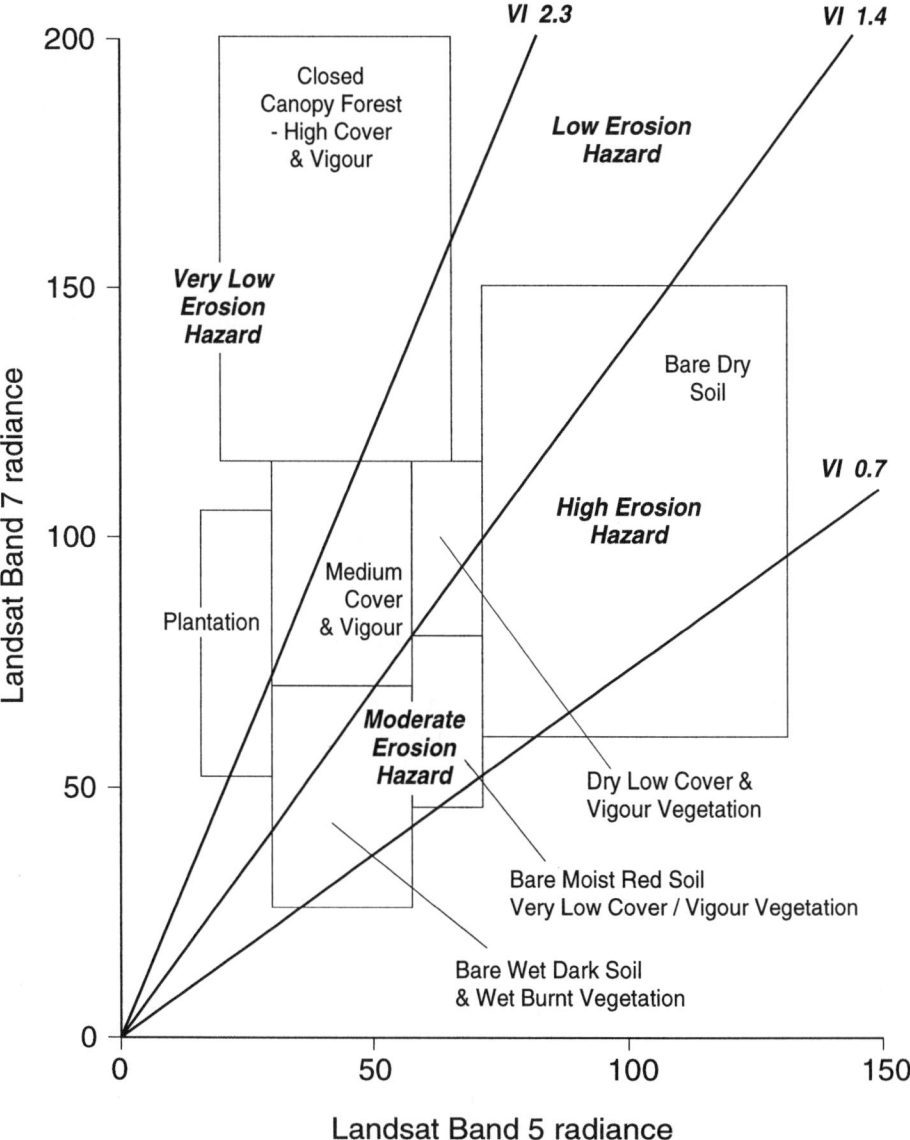

Figure 8.5 Cover assemblages and erosion classes for lower Shire valley, Malawi. VI = vegetation index values.

Results

For the lower Shire valley extract for December (Plate 2), the box classification results are shown in Plate 3. The classes are then regrouped into erosion
hazard ratings (Plate 4). The accuracy of the classification was measured by
comparing 30 samples from each land cover type derived from the imagery
with test sites on the 1:50 000 maps to give a qualitative impression of the
relative accuracies. This was consistently high for the classification in question (Plate 4).

Whilst vegetation indices and percentage vegetation cover are strongly
correlated, it is not possible to use the ratioed values as a direct measure of
percentage vegetation cover. The relationship is non-linear and varies with
changes in soil and cover type. As a result, slight changes were made in
assigning erosion hazard ratings to cover types. The most dense and vigor-
ously growing, medium-cover group was given a very low hazard rating,
whilst the least dense, medium-cover and medium-vigour group was allocated to the moderate hazard group. Dry, low-vigour and low-cover vegetation was given a severe hazard rating.

Erosion hazard classes derived from vegetation indexing (Plate 5) appear
to be more homogeneous than those from box classification. Boundaries for
marshes and irrigated crops were much more distinct, and greater discrimination was possible in the medium-cover/medium-vigour groupings in
open canopy savanna, with appropriate differences in allocation to hazard
rating classes. Bare wet dark soil was still assigned to an inappropriate
erosion hazard rating, as was also the case using the box classification.
However, vegetation indexing generally separates cover types more effectively than the box classification. This generalization, based on analysis of
December imagery, is not applicable to imagery acquired in September at the
height of the dry season, as widespread dehydration of the vegetation
produces poor inter-band correlation.

Combinations of Soil Erosion Factors

Data from Figure 8.2 have been combined to give a map of overall erosion
hazard based on the three factors of rainfall erosivity, soil group and slope
(Figure 8.6). Numerical values of 1 (low) to 4 (high) were used for each factor.
The final scores were rescaled to give four overall degrees of erosion hazard for
visual comparison with the vegetation index classification. This combination
procedure is similar to that used by Stocking and Elwell (1976), and assumes all
erosion factors are of equal importance in hazard assessment. This assumption
is acceptable in view of the uncertain nature of some of the data and the level of

understanding of soil erosion processes in this environment. A more detailed analysis, however, should consider factor weighting.

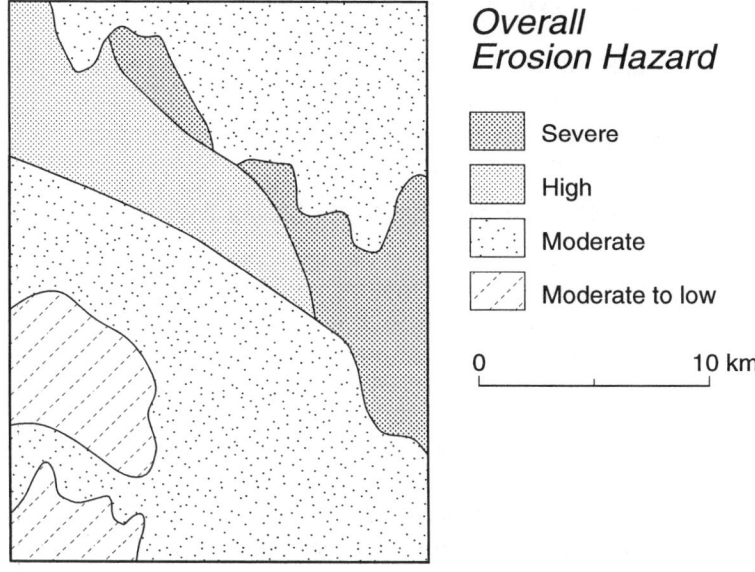

Figure 8.6 Overall erosion hazard for lower Shire valley, Malawi.

Conclusions

Problems and Potentials of the Technique

The box classifier developed in this study separates most land cover classes. The classes present significantly different soil erosion hazards. The spectral inseparability of wet burnt vegetation and bare wet dark soil has not been overcome (Figure 8.4). This causes an underestimation of the erosion hazard presented by the latter. The severity of the problem depends upon the spatial extent of the classes concerned. Vegetation index thresholding separates land cover types of different erosion hazard more effectively. However, wet burnt vegetation and bare wet dark soil remain inseparable, as do senescent vegetation and other medium-cover and medium-vigour vegetation types. For example, senescent natural grassland and maize stubble may be confused. When the stubble is cleared for cultivation, the erosion hazard may be underestimated.

Aerial photography contemporaneous with the Landsat imagery, and a programme of ground data collection of vegetation cover variables would allow a more accurate assignment of hazard ratings and the testing of alternate classifiers, like the maximum likelihood classifier. Principal components analysis may warrant investigation, as the second principal component is related to phenological growth of vegetation (Lodwick, 1979).

Future Developments for Soil Erosion Hazard Monitoring

After evaluation of imagery from June, September and December, it is considered that imagery acquired two to three weeks after the onset of the wet season in mid-November is the most useful for assessing erosion hazard. The vegetation cover in early December is critical for hazard assessment as this is the period of maximum erosion (Figure 8.1). Vegetation at this time exhibits a flush of growth which produces a wide dynamic range in digital values that makes for much easier cover type discrimination than in June or September. In early December, the cultivated land has been cleared and planted, but crop and weed growth is limited. In atmospheric terms, early December has a lower cloud cover than later in the wet season, dust content is relatively low, and the high sun angle minimizes topographic shadowing.

Of the techniques considered, vegetation indexing has advantages for the assessment of vegetation cover. The combination of rainfall erosivity, soil erodability, slope values and vegetation cover is best achieved by digital means if a suitable Geographic Information System is available.

The weighting of erosion factors requires further investigation. Those factors which are considered to be more important or have greater variability should be weighted to achieve a greater influence in the overall hazard rating. Soil erodability is considered more important than slope, and both are considered more important than rainfall erosivity and should be weighted accordingly. Vegetation cover is probably the most important and most variable factor, and should be assigned the highest weighting of all four variables. A detection technique is also required for monitoring both seasonal and annual changes in vegetation cover. With imagery geometrically rectified to the same base, vegetation indices may then be differenced to identify areas of change (Nelson, 1983).

Finally, some impression of soil erosion hazard may be obtained from visual inspection of a contrast-stretched false colour composite, given some prior experience with Landsat imagery. Many cover types can be related to different soil erosion hazards and they are distinctive on December imagery (Plate 2). Visual interpretation can provide a useful starting point for direct ground information gathering and a preliminary to more rigorous digital analysis.

Acknowledgements

Thanks are due to Jennifer Smith who initiated this research, to the Macaulay Land Use Research Institute, Aberdeen for image-processing facilities, and to Graeme Sandeman and Jim Allan of the School of Geography and Geology, University of St Andrews who produced the illustrations.

References

Anderson, J.R., Hardy, E.E., Roach, J.T. and Witmer, R.E. (1976) A land use and land cover classification system for use with remote sensor data. *United States Geological Survey Professional Paper*, **964**, 1–28.

Badcock, W.J. (1949) Soil conservation in Nyasaland. *Carona*, **1** (8), 15–16.

Brown, P. and Young, A. (1965) *The Physical Environment of Central Malawi*. Zomba: Department of Agriculture.

Directorate of Overseas Surveys (DOS) (1965) *Malawi: Natural Regions and Areas Maps*. Tolworth: Directorate of Overseas Surveys.

Douglas, M.G. (1988) Integrating conservation into farming systems: the Malawi experience. In W.C. Moldenhauer and N.W. Hudson (eds), *Conservation Farming on Steep Lands*. Ankeny: Soil and Water Conservation Society, pp. 215–27.

Elwell, M. and Stocking, M. (1976) Vegetal cover to estimate soil erosion hazard in Rhodesia. *Geoderma*, **15**, 61–70.

FAO (1981) *Tropical Forest Resources Assessment Project – Forest Reserves in Tropical Africa*. Rome: Food and Agriculture Organization.

Fournier, P. (1960) *Climat et érosion: la relation entre l'érosion du sol par l'eau et les précipitations atmosphériques*. Paris: Presses Universitaires de France.

Gratz, R.D. and Gentle, M.R. (1982) The relationships between reflectance in the Landsat wavebands and the composition of an Australian semi-arid shrub rangeland. *Photogrammetric Engineering and Remote Sensing*, **48**, 1721–30.

Green, J. and Soulsby, J.A. (eds) (1983) *National Atlas of Malawi*. Blantyre: Department of Surveys.

Lillesand, T.M. and Kiefer, R.W. (1987) *Remote Sensing and Image Interpretation*. New York: John Wiley and Sons.

Lodwick, G.D. (1979) Measuring ecological change in multi-temporal Landsat data using principal components. *Proceedings of 13th International Symposium on Remote Sensing of the Environment*. Michigan: Ann Arbor, pp. 1131–9.

Mitchell, J.K. and Bubenzer, G.D. (1980) Soil loss estimation. In M.J. Kirkby and R.P.C. Morgan (eds), *Soil Erosion*. London: John Wiley and Sons, pp. 17–62.

National Statistical Office (1970) *National Sample Survey of Agriculture 1968–69*. Zomba: National Statistical Office.

Nelson, R.F. (1983) Detecting forest canopy change due to insect activity using Landsat MSS. *Photogrammetric Engineering and Remote Sensing*, **49**, 1303–14.

O'Keefe, P. and Munslow, B. (eds) (1984) *Energy and Development in Southern Africa*. Stockholm: Biejer Institute.

Roose, E.J. (1975) *Erosion et ruissellement en Afrique de l'Ouest*. Abidjan: Office de la Recherche Scientifique et Technique Outre-Mer.

Soulsby, J.A. (1992) Environmental pressures and development – a framework for development in East Africa. *Zeszyty Naukowe Uniwersytetu Jagiellonskiego Prace Geografiiczne*, **88**, 17–24.

Stocking, M. and Elwell, M. (1973) Soil erosion hazard in Rhodesia. *Rhodesia Agricultural Journal*, **70** (4), 93–101.

Stocking, M. and Elwell, M. (1976) Rainfall erosivity over Rhodesia. *Transactions of Institute of British Geographers*, N.S. 1, 231–45.

Tucker, C.J. (1979) Red and photographic infra-red linear combinations for monitoring vegetation. *Remote Sensing of the Environment*, **8**, 127–50.

Weil, R.R. (1982) Maize-weed competition and soil erosion in unweeded maize. *Tropical Agriculture*, **59**, 207–13.

World Resources Institute (1992) *World Resources 1992–93*. Oxford: Oxford University Press.

Young, A. and Brown, P. (1961) *The Physical Environment of Northern Nyasaland*. Zomba: Government Printer.

9

Resettlement and the Management of Environmental Degradation in the African Farming Areas of Zimbabwe

Jennifer Elliott

The resettlement programme in Zimbabwe was initiated shortly after independence in 1980. By 1991, 52 000 families had been voluntarily relocated, largely from the African (communal) farming areas, to 3.3 million ha of former European-owned commercial farmlands. The programme was established with a diversity of objectives, reflecting both the immediate and longer-term tasks of the independent government. These included the need to provide groups such as ex-combatants, displaced persons and landless people with the means to a livelihood and also the government's commitment to a socialist transformation based on 'growth with equity'. Consistent throughout the fifteen-year history of the programme has been the intention that resettlement should assist in the alleviation of population pressure and environmental degradation in the communal areas.

This chapter reviews the changes in resettlement policy and procedures in Zimbabwe and assesses the impact of resettlement on neighbouring environments as evidenced in two case study sites.

Human Resource Relationships in Rural Zimbabwe

On independence in Zimbabwe, the legacy of inequitable policies of land apportionment during the colonial period was evident in the wide disparity in access to land according to racial group. For example, in 1982, rural Africans constituted 59 per cent of the population and had access to 46 per cent of the land area, much of it in less fertile regions. The average population density of the communal areas in 1986 was 25.5 persons km^{-2} in comparison to 7.6 in the 'general lands' that encompassed the large-scale commercial farming sector (Zinyama and Whitlow, 1986). In addition, the independent government inherited a dual system of administration and development for the European and African areas, within which colonial conceptions of, and responses to, problems of population, environment and development had differed markedly (Beinart, 1984; Phimister, 1986; Elliott, 1991).

Early policy documents attributed the low level of socio-economic development and evident problems of environmental degradation in the communal areas to a set of linked agro-ecological, socio-economic, infrastructural and institutional problems inherited from the past. Speaking to the Natural Resources Board shortly after his inauguration as Prime Minister, Mugabe suggested that the communal areas had been 'seriously neglected over the years in every way by successive governments which has led to the very poor state of conservation and productivity generally' (Mugabe, 1980, p. 288). He committed his government to establishing a unitary system of administration across all sectors, the rapid extension of conservation institutions, extension advice and resources to the communal areas, and a programme of population movement through resettlement.

Resettlement commenced in late 1980. The initial plan was to resettle 17 500 families on 1.1 million ha of land over a five-year period. In 1982, the target was raised dramatically to include 162 000 households on 10 million ha of land over the same period (Republic of Zimbabwe, 1982). In 1989, the National Land Policy established a ceiling on land transfers of 8.3 million ha, equivalent to 25 per cent of total farmland (Government of Zimbabwe, 1989). Three 'models' of resettlement were established: model A based on individual family farms with settlers living in nucleated villages and sharing grazing land; model B involving co-operative farms organized on a collective basis; and model C encompassing a system of core estates and outgrowers. In practice, model A settlement has accounted for over 80 per cent of total expenditure to 1987 (Cusworth and Walker, 1988).

In 1992, a review of existing schemes led to some modifications and new models, including several model A variants according to local conditions, limits on future resettlement under model B, and a new model D based on livestock production and suited to areas of low rainfall. Selected settlers are allocated three permits: to cultivate the land, to hold and graze stock, and to reside in the scheme area. Permit holders are required to 'permanently and personally reside' on the land allocated to them, i.e. to be engaged in full-time farming and to renounce all rights to cultivate land or hold and graze stock in any communal area.

Initially, it was decreed that, wherever possible, land for resettlement would be acquired adjacent to communal areas suffering acute population pressure and environmental degradation (Government of Zimbabwe, 1981). In practice, however, the location and pace of land acquisition were restricted by the private property provisions of the Lancaster House agreement, which limited transfers to a 'willing-seller / willing-buyer' basis. Nor was the selection of farmers in any way linked to the population pressure of particular communal areas. Settlers were selected according to individual need, with priority being given to the landless and displaced.

In 1985, the objective of alleviating population pressure in the communal areas was strengthened through the integration of communal land reorganization with the resettlement process. It was envisaged that people displaced through rationalization of land use in these areas would become eligible for resettlement. However, the policy was not implemented, and, in 1992, it was concluded that this was partly due to the lack of a coherent policy and procedural framework. A more detailed framework was consequently established in respect of both communal land reorganization and the translocation resettlement programme with a view to being 'implemented concurrently' (DERUDE, 1992). The National Land Policy is intended to provide the required coherence in rural development policy covering resettlement and communal and commercial farming areas, whilst the Land Acquisition Bill (1992) enables the compulsory purchase of land for the resettlement programme.

In summary, although the objectives of the resettlement programme are varied, the intention that movement of people from the communal areas should alleviate population pressure and environmental degradation therein has been consistent throughout three major policy revisions. Several constraints to achieving this objective have been identified at a national policy level and modifications undertaken to enhance the prospects for development in the communal areas. Unfortunately, however, objective information on the local impact of resettlement has been very limited. The following sections are based on field research in two communal areas from which families have been resettled. The research identifies a number of interrelated local factors, physical and human, that have significantly affected the impact of resettlement on resource use in the communal areas from which settlers were drawn.

Insights from the Case Studies

The following results are part of a wider research project undertaken in 1992–93 into the sustainability of household responses to fuelwood needs in the resettlement areas of Zimbabwe. A survey of 95 households in the villages of Nhowe (Svosve communal area) and Nhema (Shurugwi communal area) was carried out, focusing on environmental and social change since movement from these villages to the Wenimbi and Tokwe resettlement schemes began in 1985 (Figure 9.1).

Relief of Population Pressure

At the national level, population movement through the resettlement programme was constrained, at least until 1990, by the limitations on land acquisition set by the Lancaster House agreement. Of the total families moved within the programme, it is estimated that 85 per cent have come from the communal areas (MLARR, 1986). At this level, the programme at its current pace cannot be expected to alleviate population pressure substantially in those areas, where the average population growth exceeds 3 per cent *per annum* (CSO, 1992). Such aggregate statistics may disguise greater benefits at the local level, but, as Table 9.1 highlights, resettlement to the Tokwe and Wenimbi schemes has not outpaced population growth in the neighbouring communal areas. The largest impact of resettlement, however, has been in the smaller communal area of Svosve, which also had the highest population density in 1982. Again, this level of analysis may obscure the experience of particular villages from which settlers have been drawn.

For the communal households studied, population pressure was most evident in terms of access to arable land. When asked to consider changes in their village areas since independence, 88 per cent of respondents suggested that the resettlement programme had had no real impact on land availability. As one Svosve resident stated, 'resettlement only works for those who have moved'. For 21 per cent of the respondents, the limited impact was attributed to the fact that few of the people moved were actually from the communal areas under study. A further 16 per cent perceived that the demand for land created by young adults within the area had outpaced movements to the resettlement scheme. However, natural increase was not the only cause of population growth and land availability; 22 per cent of respondents blamed the arrival of people from other regions for the limited impact of the resettlement programme on local population pressure. In Zimbabwe, a significant proportion of labour on European farms has habitually come from outside the country. Although it has always been policy that 'non-indigenous natives' should return to their home country at the end of their employment or their ability to work, many have not done so for economic and social reasons. Indeed, some have long-established contacts, including marriage, with communal area people. It is thus likely that many of them will have chosen to settle in the communal areas when commercial farms were purchased for resettlement.

The precise extent of this 'infiltration' is not known, but its qualitative aspects may be as important as its quantitative ones. Particularly in Svosve,

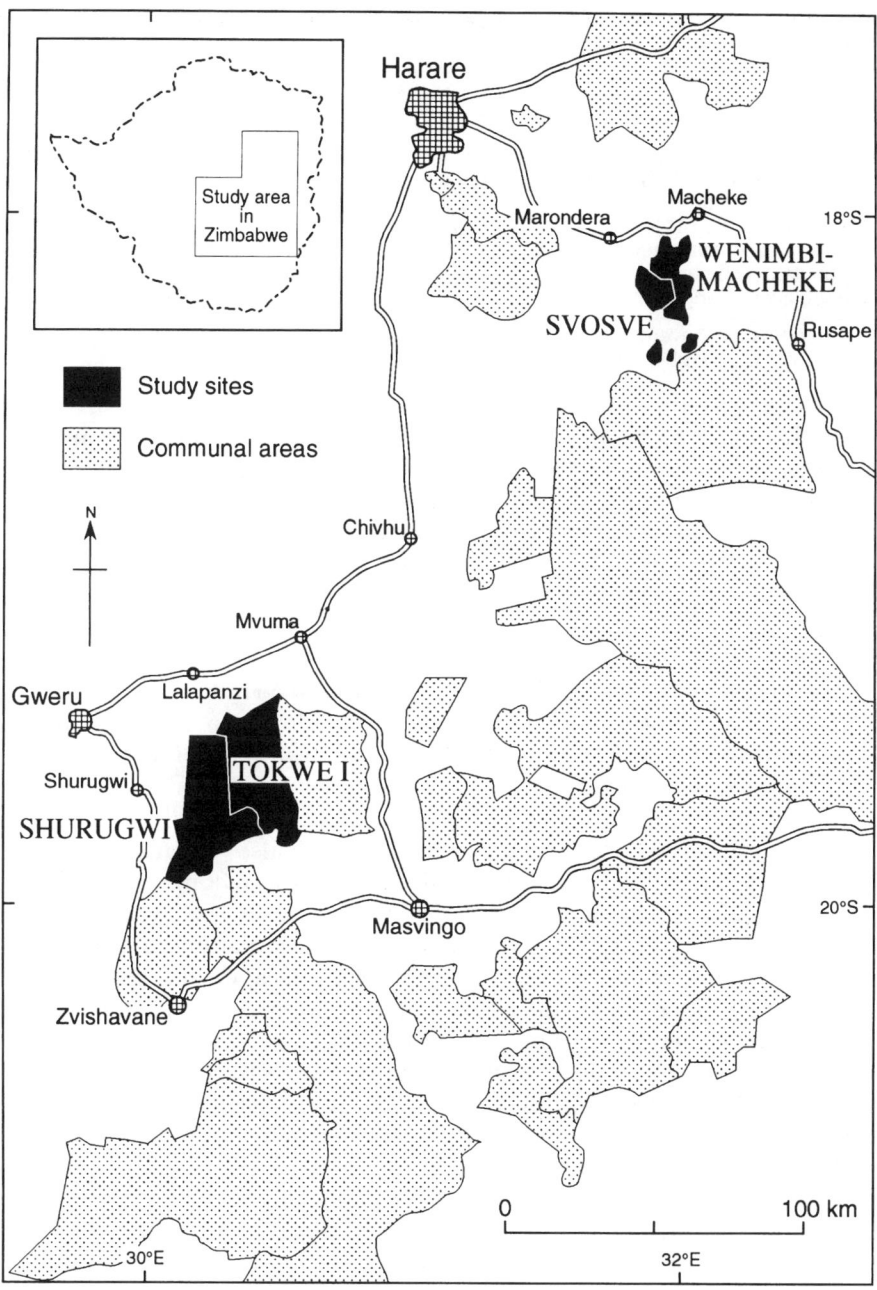

Figure 9.1 Location of case study sites in Zimbabwe.

Table 9.1 Aspects of population change in Svosve and Shurugwi communal areas in 1982–1992.

	Svosve	Shurugwi
Area (km²)	110	825
Population in 1982	5558	41 021
Population density in 1982 (persons km⁻²)	50.5	49.7
Families resettled by 1992	480	177
Persons resettled by 1992*	2304	850
Natural increase, 1982–1992*	2007	14 806
Net population change	+ 297	+ 13 956
Net population change (%)	5.3	34.0

* Based on national average family size and rates of population growth.
Source: after CSO (1984, 1992), field survey.

where the presence of outsiders was most widespread, it was not so much the numbers of people *per se* that was perceived as problematic as the fact that they were 'not of our people' and caused the breakdown of community structures and societal norms of behaviour. One Svosve resident was adamant that 'the people who replaced those who were resettled are not from Svosve and hence there is little understanding between villagers. Foreigners have no respect for norms and cultural values and therefore life is more difficult.'

Population densities, as in Table 9.1, clearly give no indication of the range of social, political and economic factors combining to influence human resource relationships in an area. Furthermore, the impact of resettlement on population pressure in the communal areas varies not only with the numbers involved in the programme, but also with who moves and the extent of the resources which they used prior to resettlement. It also depends on whether the resources are fully or partially reallocated and to whom.

The lack of formal procedures for the disposal of land vacated in the communal areas has limited the impact of the programme on population pressure at the national level (MLARR, 1992). Despite the obligation on settlers to renounce traditional rights within communal areas, this has been considered an 'alien if not impossible practice for many people for social and cultural reasons' (MLARR, 1986, p. 24) and violations have been widespread. A total of 21 per cent of respondents from the two villages perceived the reallocation of lands by settlers within their own families as the reason for the limited impact of resettlement on population pressure. In addition, there was evidence of an emerging private market in land transfers. Several respondents reported difficulties in accessing their rights to land in the absence of payment to the village headman, who often has *de facto* control over land allocation. The drought and retrenchment from industry were perceived to be factors causing a return of people to the communal areas and a willingness to pay village heads for permission to settle. In the Svosve area, outsiders from neighbouring commercial farms were also involved in this practice.

In summary, survey data confirm that aspects of the resettlement policy itself have limited the capacity of the programme to alleviate population pressure in the communal areas. The analysis has also identified new factors, such as who is moved from the communal areas and who is resettled in neighbouring villages, as important determinants of the impact of resettlement on population pressure in the communal areas.

Alleviation of Environmental Degradation

National surveys have highlighted the strong spatial continuities between environmental degradation and land tenure in Zimbabwe. In terms of erosion, for example, 27 per cent of the communal areas suffer 'severe' to 'very severe' levels, as compared to only 1.6 per cent in the case of commercial farmlands (Whitlow, 1988). As indicated, one of the objectives of the resettlement programme has been to create a degree of 'elbow room' in the communal areas and to allow conservation and development of resources therein. However, increased productivity on communal area farms – for example, a twelve-fold increase of maize and sorghum sold through official marketing boards between 1980/81 and 1985/86 has been referred to as Africa's agricultural success story (Cliffe, 1988) and has even been cited as justification for terminating the resettlement programme.

Since the initiation of resettlement in the study sites, the average livestock holding per family in the two villages has declined from 9.0 to 6.6 head of cattle. Likewise, the average area cultivated has fallen from 1.7 to 1.5 ha. In addition, respondents highlighted persisting environmental problems such as water supply (38 per cent), depletion of grazing resources (28 per cent), declining soil fertility (18 per cent) and land shortage (7 per cent). Environmental decline was also evident in the greater distances now travelled in fuelwood collection. Whereas 81 per cent of respondents reported that fuelwood was 'easy' or 'fairly easy' to obtain ten years ago, 64 per cent perceived the situation today as 'hard' or 'very hard'.

Interrelated factors limit the impact of the resettlement programme on resource use and degradation in the communal areas. For example, it is evident that the resource base on which villagers depend has never been solely contained within the administrative boundaries of the communal area. The resettlement programme, however, has led to significant changes in the nature and extent of interactions across boundaries of one tenure type and another. In addition, the population changes due to resettlement are perceived by respondents to affect many aspects of resource use within the communal areas.

Whilst rural development and planning, both pre- and post-independence, have generally treated communal and commercial farming areas as distinct, the case studies show that the patterns and processes of natural resource use cut across the boundaries set by land ownership. In the extreme, this is illustrated by several respondents who asserted that 'we are now independent and the land belongs to all people'. The mechanisms for individual access to resources, particularly of grazing and woodlands, during European ownership appear to have been diverse; respondents reported access being granted variously, at particular times, in particular locations, and in return for labour and/or cash. Access was not enjoyed by all villagers and such continues to be the case.

Fuelwood collection exemplifies the interactions between communal residents and neighbouring resettlement areas. The extent of collection beyond the communal area boundaries for the case study sites is shown in Table 9.2. Respondents were asked to identify the location of their main source, past and present, for major wood-using activities. It is clear that wood collection has generally increased since the conversion of neighbouring commercial farmlands to resettlement villages. The biggest change since independence for one Svosve resident was that 'we now have the freedom to enter the farms which we were not allowed when they were under white farmers'. Yet 66 per cent of the sample perceived that it was now harder to collect wood than it was ten years ago, 28 per cent attributing this to deteriorating access to the resettlement scheme areas.

Table 9.2 Percentage of respondents collecting fuelwood in the resettlement area.

Purpose	Early 1980s	1991/92
Cooking	36	59
Construction	31	44
Brick-making	23	40
Brewing	21	38

Various factors are important in differentiating local experience of access to and conflicts over the use of woodland resources from neighbouring resettlement scheme areas. They include the stance taken by local officers over 'illegal' incursions, the local ecologies of the source area and the social composition of the settler village. Local officers in Shurugwi, for example, were reluctant to take action against communal or resettlement farmers engaged in collecting or selling wood, in recognition of the total lack of alternative fuelwood options. In contrast, in Svosve, there have been several prosecutions and there are outstanding disputes between particular villages and the neighbouring resettlement farmers. For the communal area respondents, it was not the change of ownership *per se*, but the movement of people unknown to them which was curtailing their access to resources. Equally, the settlement of local people was not enough to ensure access for all communal residents; some respondents suggested that access to grazing and woodland resources in the resettlement areas was now enjoyed only by the relatives and friends of those moved. Others suggested that access had deteriorated for everyone due to changing attitudes on the part of settlers: 'now the resettled people behave as if they do not know people from the communal area'.

The nature and extent of resource use by communal-area residents within the resettlement areas clearly have implications for the communal areas themselves. Several respondents in Shurugwi, for example, linked an improvement in grassland in their village directly to the change in ownership of the neighbouring land and their new ability to graze their cattle in the resettlement area 'since these people are our children'. However, population changes in the communal areas were also perceived to influence many aspects of resource use therein. For example, in both the villages under study, references were made to the continued shortage of arable lands which caused the allocation of new lands in locations unsuited to cultivation. Respondents acknowledged that the recipients were entitled to lands within the village, but the location in established grazing areas, for example, was perceived to be a problem for the environment and their daily activities. One Shurugwi resident stated that 'people are just doing what they want – they are settling where they like and there are no rules governing the movement of people into the grazing areas'. Many attributed their need to collect wood in resettlement areas as being due to the recent cultivation and associated clearance of trees in the grazing areas. The same problem was voiced by a Svosve resident: 'what used to be fuelwood sources are now peoples' fields'. Several respondents doubted the ability of the community to implement grazing schemes whilst such people occupied the land. In Svosve, the additional issue of the allocation of land to outsiders compounded the concern of some respondents.

The importance of perceived 'social cohesiveness' in the management of resources was evident in these interviews. As one Svosve resident stated,

> there are no longer restrictions on what species are to be felled and which ones should not ... there is only one well and it is drying up because foreigners no longer follow the cultural norms and rituals and hence the spirits are angered, which leads to the drying up of the well.

This accords with a previous study of institutional control over environmental resources:

> It is often difficult for villagers to exercise the same control over new migrants, particularly those from a different tradition, that they exercise over those socialised in the village. Thus traditional controls begin to erode with the introduction of new residents, and sometimes, new religious beliefs. (Nhira and Fortmann, 1993, p. 149)

Conclusion

The determinants of agrarian strategy at any one place or time are diverse. This chapter documents the changes in resettlement policy in Zimbabwe as they have affected resource management in the communal areas. Even though the political–economic structures and interests that have supported and shaped policy changes have not been analysed here, it is evident that the technical and economic aspects of resettlement have been prioritized at some cost to the social and environmental factors. The chapter also shows how the limitations set by national policies and procedures of resettlement have been transformed locally to impact on population pressure and resource use. In addition, by focusing on woodlands in the case study areas, local forces have been identified that further influence population pressure and resource use.

The impact of recent changes in resettlement policy, such as the alignment of translocation resettlement with the planned programme of communal area reorganization or the new settler selection criteria, has yet to be seen in practice. However, the present case studies confirm that resource use within the communal areas is indeed linked to the resettlement programme. Long-established 'resource-sharing' practices, for example, are one of the inter-actions that exist across land tenure boundaries and which have been modified with the conversion of farms to resettlement schemes. Environmental change is seen by communal-area respondents to be linked to social change within the village and between themselves and neighbouring settlers, as well as to political and institutional factors. The substantial diversity in the experience of population and environmental change at the local level suggests the value of further case study work in understanding the prospective achievements of the resettlement policy and in formulating flexible solutions for resource management in the communal areas of Zimbabwe.

Acknowledgement

This paper draws on field research funded by the UK Overseas Development Administration as part of their Population and Environment Research Programme.

References

Beinart, W. (1984) Soil erosion, conservationism and ideas about development: a Southern African exploration, 1900–1960. *Journal of Southern African Studies*, **11**, 52–83.

Cliffe, L. (1988) Zimbabwe's agricultural 'success' and food security in Southern Africa. *Review of African Political Economy*, **43**, 4–25.

CSO (1984) *1982 Population Census: A Preliminary Assessment*. Harare: Central Statistics Office.

CSO (1992) *Zimbabwe Census 1992. Preliminary Report*. Harare: Central Statistics Office.

Cusworth, J. and Walker, J. (1988) *Land Resettlement in Zimbabwe: A Preliminary Evaluation*. London: Overseas Development Administration Evaluation Report EV434.

DERUDE (1992) *Policies and Procedures Resettlement and the Reorganisation and Development of the Communal Lands*. Harare: Department of Rural and Urban Development, Ministry of Local Government, Rural and Urban Development.

Elliott, J.A. (1991) Environmental degradation, soil conservation and the colonial and post-colonial state in Rhodesia/Zimbabwe. In C. Dixon and M. Heffernan (eds), *Colonialism and Development in the Contemporary World*. London: Mansell, pp. 72–92.

Government of Zimbabwe (1981) *Resettlement Programme: Policies and Procedures*. Harare: Ministry of Lands, Resettlement and Rural Development.

Government of Zimbabwe (1989) *National Land Policy*. Harare: Ministry of Lands, Agriculture and Rural Resettlement.

MLARR (1986) *A Sample Survey of Settler Households in Normal Intensive Model A Resettlement Schemes*. Harare: Monitoring and Evaluation Unit, Ministry of Lands, Agriculture and Rural Resettlement.

MLARR (1992) *Second Report of Settler Households in Normal Intensive Model A Resettlement Scheme*. Harare: Monitoring and Evaluation Unit, Ministry of Lands, Agriculture and Rural Resettlement.

Mugabe, R. (1980) Opening address of the First National Conservation Congress of Zimbabwe. *Zimbabwe Science News*, **14**, 287–9.

Nhira, C. and Fortmann, L. (1993) Local woodland management: realities at the grass roots. In P.N. Bradley and K. McNamara (eds), *Living with Trees. Policies for Forestry Management in Zimbabwe*. World Bank Technical Paper No. 210, pp. 139–55.

Phimister, I. (1986) Discourse and the discipline of historical context: conservationism and ideas about development in Southern Rhodesia 1930–1950. *Journal of South African Studies*, **5**, 263–75.

Republic of Zimbabwe (1982) *Transitional National Development Plan 1982–1985*. Harare: Ministry of Finance.

Whitlow, J.R. (1988) *Land Degradation in Zimbabwe: A Geographical Study*. Harare: Department of Natural Resources/University of Zimbabwe.

Zinyama, L. and Whitlow, R. (1986) Changing patterns of population distribution in Zimbabwe. *GeoJournal*, **13**, 365–84.

10

Environmental Problems in Kenya's Arid and Semi-arid Lands

Michael B.K Darkoh

The arid and semi-arid lands of Kenya cover approximately 506 000 km², or 88 per cent of the land area of the country (Figure 10.1). Although the arid and semi-arid lands (ASAL) are often treated as a single unit, they span four agro-ecological zones and significant differences exist from one area to another (Table 10.1). Approximately 70 per cent of the ASAL (352 000 km²) is truly arid (zones VI and VII). Conditions in these drier and poorer areas bear little relationship to conditions in the wetter and more productive areas (zones IV and V).

Important as these differences are, however, the ASAL share a number of problems. They are poor in natural resources, and, by virtue of their low population and inadequate social and educational services, they have been considered of little importance in political terms. In addition, they have failed to articulate demands for increased development as vociferously as other more dynamic areas of Kenya. Meanwhile, several ASAL areas have received land-hungry migrants from the densely populated Kenyan highlands, which has raised their population growth rate above the Kenyan average. As a result, the ability of agriculture to sustain the rural population of the ASAL is declining because many households have insufficient land and/or livestock to meet subsistence requirements (Helland, 1987). This situation is exacerbated in the wetter areas by the subdivision of land holdings and conversion of grazing lands to marginal cropping, and in the drier areas by overgrazing, deforestation and soil erosion. Formal sector jobs are scarce and expensive to create in the ASAL, and an increasing number of people in both agricultural and pastoral areas now depend on non-agricultural income for subsistence and survival. The ASAL have always been vulnerable to periodic drought, with the driest areas suffering the least reliable rainfall. Recent intensification

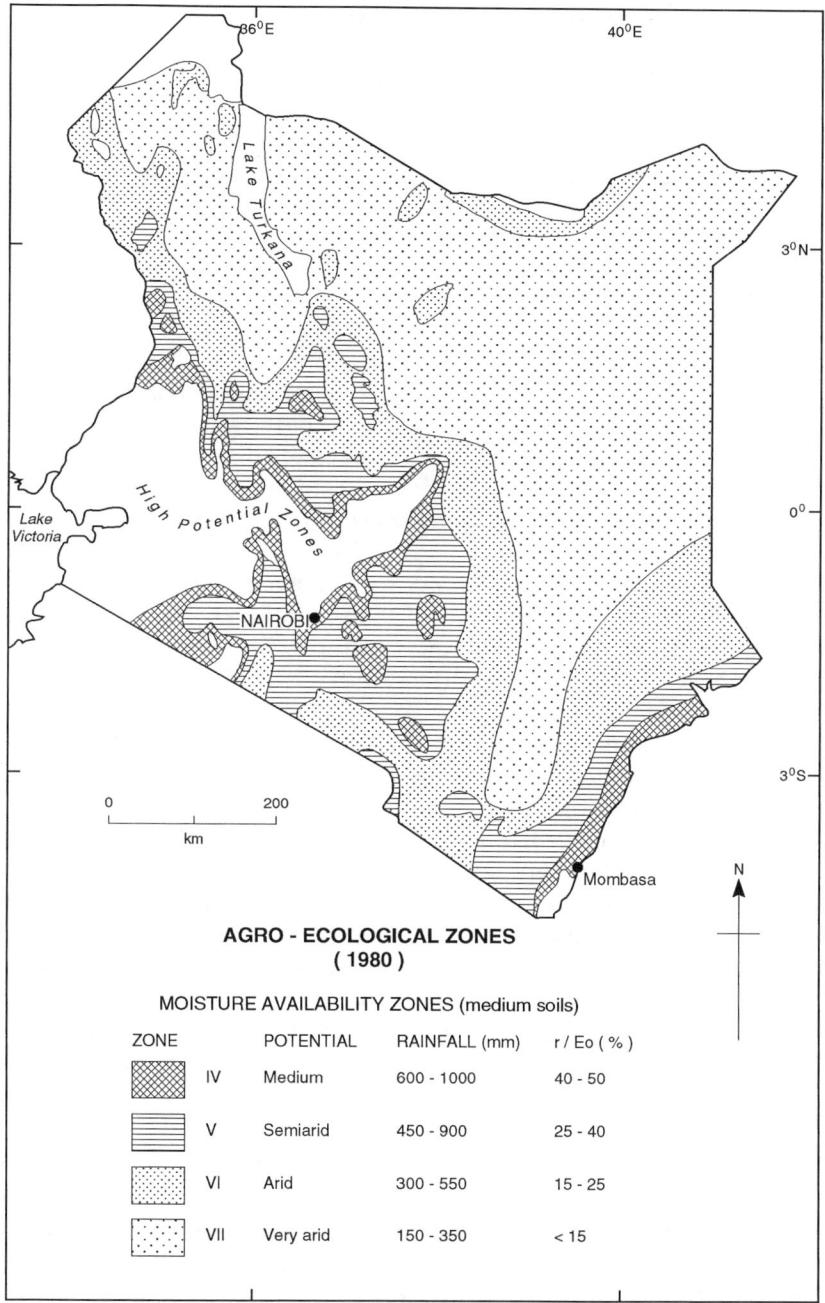

AGRO - ECOLOGICAL ZONES
(1980)

MOISTURE AVAILABILITY ZONES (medium soils)

ZONE		POTENTIAL	RAINFALL (mm)	r / Eo (%)
	IV	Medium	600 - 1000	40 - 50
	V	Semiarid	450 - 900	25 - 40
	VI	Arid	300 - 550	15 - 25
	VII	Very arid	150 - 350	< 15

Figure 10.1 Arid and semi-arid lands in Kenya.

Table 10.1 Area of agro-ecological zones (AEZ) in Kenya.

Zone	Description	Percentage r/EO*	Area (km²)	Percentage land area of Kenya
IV	Semi-humid	40–50	27 000	5
V	Semi-arid	25–40	87 000	15
VI	Arid	15–25	126 000	22
VII	Very arid	15	226 000	46
			506 000	88

* r/EO = rainfall/evapotranspiration ratio
Source: after Jaetzold and Schmidt (1982).

of land use, particularly in marginal agricultural areas, has increased the risks of large-scale land degradation and general impoverishment.

In the drier parts of the ASAL, nomadic pastoralism is still the main economic activity. Moisture availability severely limits opportunities for cropping, although in areas like Turkana District small-scale cultivation of sorghum and other hardy crops has traditionally been practised in more favourable sites, notably areas of impeded drainage and river floodplains. The drier areas, however, are often poorly endowed with natural resources. Land degradation may not be as acute as in more intensely exploited areas, but drought and famine are perennial features. The other important constraint on development in the drier areas is the lack of governmental attention to regional problems.

The main development issues in the ASAL are thus the related problems of poverty and food security. Famine is a constant feature of the areas and famine relief has been provided on a regular basis for several decades. In 1990, under the auspices of the World Bank and Kenya's Ministry of Reclamation and Development of Arid, Semi-arid Areas and Wastelands (MRDASW), the author was involved in the preparation of a Draft Environmental Action Plan for Kenya's ASAL and had the opportunity of touring and observing the environmental situation in 16 of the 22 (now increased to 24) administrative districts that make up the ASAL. Some of the major environmental issues and problems in the ASAL are examined here.

Land Use and the Environment

The Department of Resource Surveys and Remote Sensing (DRSRS), formerly the Kenya Rangeland Ecological Monitoring Unit, has been using satellite, aerial and ground observation to provide a continuous flow of reliable information on land use and land cover in Kenya. The land use and land cover mapping programme recently undertaken by DRSRS was based on 1:1 million scale Landsat imagery and covers the whole country. A preliminary map was produced using 1972–1980 Landsat colour composite transparencies and 1:250 000 scale base maps. According to DRSRS, the total area under cultivation in Kenya covers 91 238 km², which is 17.2 per cent of the country (Epp *et al.*, 1982). The results indicate that all land suitable for rainfed cultivation in Kenya is already in use.

Arable land for subsistence farming is in short supply in areas of high potential, resulting in the inevitable expansion of cultivation into drought-prone areas of the east and north. Detailed land-use mapping by DRSRS at 1:250 000 scale clearly demonstrates this shift with respect to the semi-arid areas of Narok and Kitui (Figures 10.2 and 10.3). Related data are provided in Table 10.2.

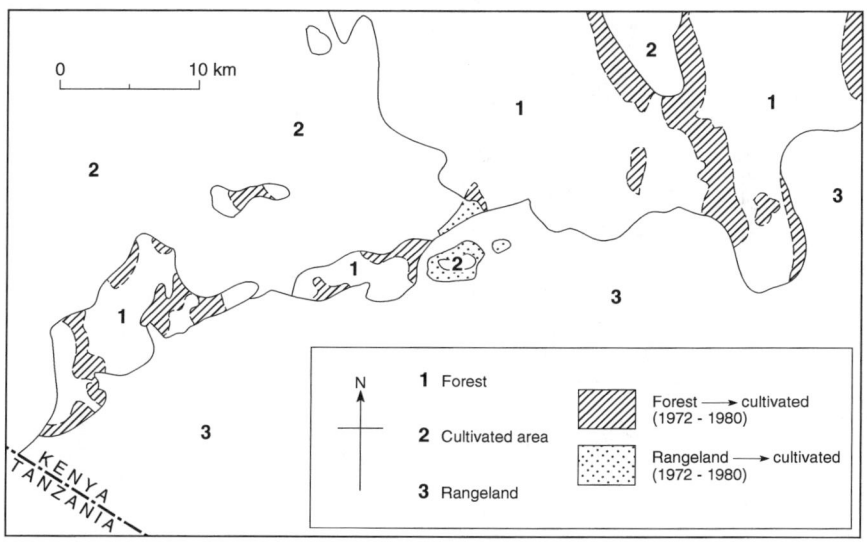

Figure 10.2 Agricultural encroachment in Narok District, 1972–80. *Source*: after Agatsiva and Mwendwa (1982).

Today, as a result of rapid population growth and the spill-over from lands of high potential to those of marginal productivity, some semi-arid districts like Kitui, Embu and Baringo are among the most threatened areas in Kenya. Their growth of population is due to both natural increase and migration (Anzagi and Bernard, 1977; Bernard and Thom, 1981; Thom and Martin, 1983). Their rangelands are overgrazed, topsoils are eroding and food short-ages are common (Bernard, 1985).

Even so, such conditions are not ubiquitous in arid and semi-arid districts undergoing rapid population growth. A more encouraging area is the Ma-chakos District to the southeast of Nairobi. The area was cited by Bernard (1985) and co-workers as one of the most threatened in Kenya, but other

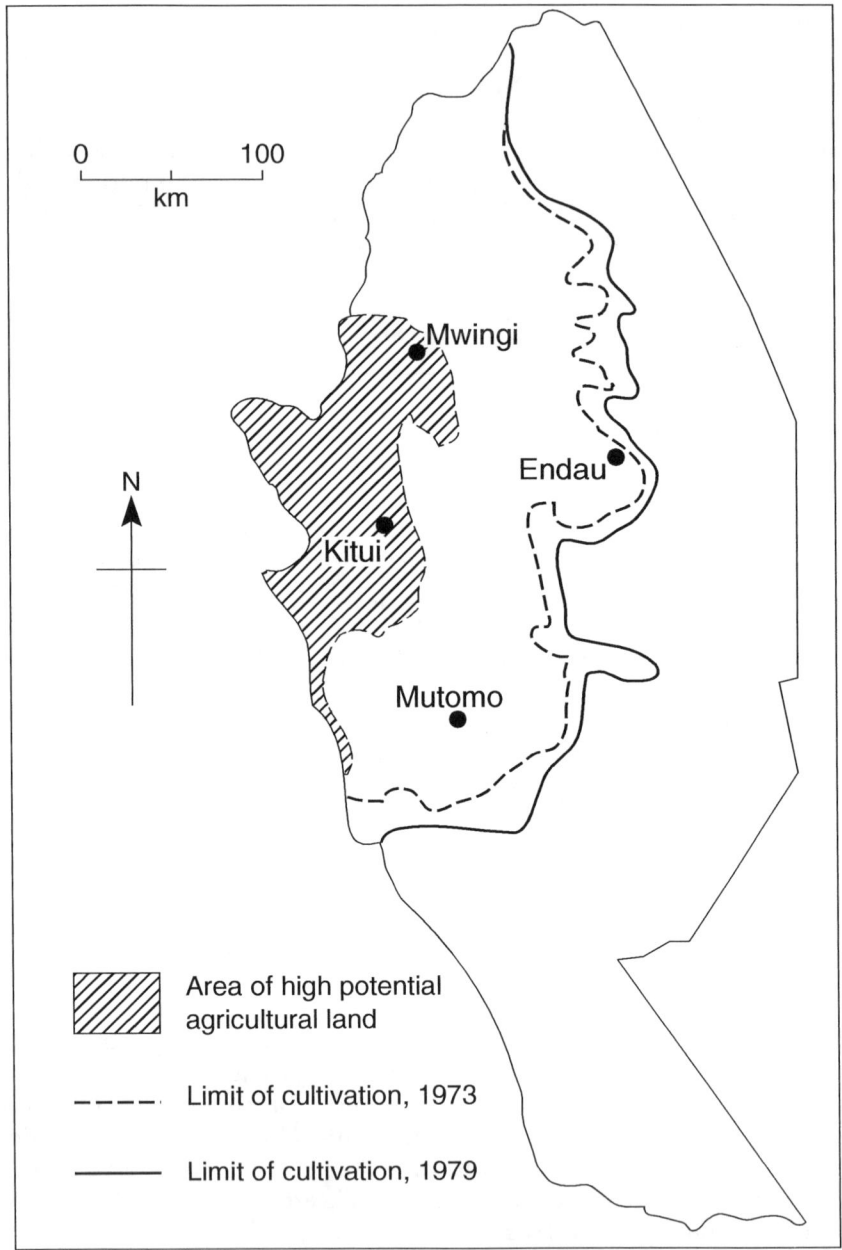

Figure 10.3 Agricultural encroachment in Kitui District, 1973–79. *Source*: after Agatsiva and Mwendwa (1982).

Table 10.2 Changes in land use in Narok and Kitui Districts.

District	Land use	Area (thousand ha)			
		1967	1973	1976	1980
Narok	Wheat cultivation	0.11	1.17	2.16	–
Kitui	Shifting cultivation	–	900	–	1080

Source: after Agatsiva and Mwendwa (1982).

investigators have questioned the linkage between local population growth and soil erosion (Barber, 1983; Tiffen, 1991, 1992; English *et al.*, 1994). Such doubts are confirmed by comprehensive time-series data gathered for the district for the period 1930–90 which show that significant development has occurred (Gichuki, 1991; Mbuvi, 1991; Mutiso *et al.*, 1991; Mortimore, 1992; Mortimore and Wellard, 1992; English *et al.*, 1994; Tiffen *et al.*, 1994).

Under colonial rule in the 1930s, the Machakos District was considered overpopulated and at risk of overexploitation. Erosion rates were high. With independence and new economic policies, however, conditions have improved. The population is still growing at more than 3 per cent *per annum*, and fallow periods continue to shorten. Yet there has been no acceleration of soil erosion. Instead improved access to technical inputs, extension services, market outlets, health care and education has led to improvements in both traditional and modern land use techniques, to cultivation of more profitable crops, and to major investments in land improvement (Mutiso, 1989; Wamalwa, 1989). Shorter fallow periods have been offset by improved soil management, and crop yields have increased faster than population (Tiffen, 1991, 1992; Hansen, 1992; English *et al.*, 1994; Tiffen *et al.*, 1994).

Migration Trends

Three types of migration are discernible in the ASAL: permanent immigration into marginal dryland farming areas; temporary or seasonal migration in search of work within and outside the ASAL; and return migration (MRDASW, 1990). Several factors influence these flows, including climate, security, tradition, the availability of central services, famine relief, employment opportunity in towns or irrigation schemes, and the seasonal labour demand in agricultural districts. With dense populations in the high-potential zones, more and more people are moving from there into marginal areas in search of farming land and employment opportunity. ASAL districts that have recently experienced high immigration include Laikipia, Narok and Kajiado.

In terms of environmental degradation, permanent immigration into marginal dryland farming areas is most problematic, as such migrants often cause land pressure and import inappropriate technologies. They also disrupt indigenous land-management systems that are based on appropriate and locally adapted technologies.

Permanent immigration and cultivation in the marginal lands of Kenya are a relatively recent phenomenon. Before colonial rule there was little movement of arable farmers into the ASAL. Traditionally, the ASAL were the home of pastoral peoples, like the Masai, Pokot, Samburu, Turkana, Galla, Rendille, Gabbra and Boran, and of abundant wildlife. Pastoralists maintained a variety of social and economic controls within their own society and between themselves and neighbouring communities. There was little competition between livestock and wildlife and little inclination on the part of neighbouring agricultural communities to move into the ASAL. The movement of farmers into the area, overstocking of the land, and famine were problems that originated during the period of colonial rule and were a direct consequence of European land alienation in the high-potential zones which created a surplus rural population in those areas (Wisner, 1977). Most land alienation occurred between 1908 and 1920. It has been estimated that by 1935 over 3.5 million ha of high-potential land in the Kenyan Highlands had been alienated by the British Colonial Government. One well-documented migration into a pastoral area took place from Kiambu to the Ngong Hills in the 1920s (HMSO, 1934). As the pressure on land continued to grow in the high-potential areas so farmers spontaneously began a search for alternative locations where agriculture could be practised. As a consequence, they moved into less favourable areas on the wetter margins of the ASAL.

More recently, movement into the ASAL has been due not only to spontaneous migration, but also to government policy that aims to alleviate land shortage in the high-potential areas by planned settlement and cultivation in the wetter margins of the ASAL. Among the planned migrations are those to government-sponsored irrigation schemes, like the Bura and Hola schemes on the Tana river. According to Campbell (1981), the areas in question, namely, the Ngong hills, Mau Marok, the Cherangani and Tugen hills, Meru, Kitui, the Chyulu hills, the Kilimanjaro foothills and the Soit Ololol escarpment, now form an interface between the sedentary farming economies of the high-potential lands and the pastoral and wildlife economies of the semi-arid rangelands.

Land Use Conflicts

In this interface zone, conflicts have arisen as a result of the intrusion of agriculture into lands traditionally used for domestic stock. There is competition for resources, particularly water, between various sectors, notably agriculture, livestock, wildlife and settlements.

One semi-arid area of recent permanent immigration is the Loitokitok area of Kajiado District (Sindiga, 1984a, 1984b; Campbell, 1986). Incoming farmers have moved primarily to the foothills of Mount Kilimanjaro and to land adjoining the perennial streams and swamps in the area. The experience of local pastoralists during the most recent drought is that they have been less able to adapt to water shortage because of the expansion of cultivation and also the gazetting of National Parks. Land use changes have also reduced access to dry-season pastures (Sindiga, 1984a, 1984b; Campbell, 1986). If current trends continue with no alternative sources of support being available, drought hazards are likely to worsen and the region will experience poverty, famine and further deterioration of both rangeland and agriculture (Bernard, 1985).

Land use conflicts have also arisen in key production areas of the ASAL. These are the riverine forests along main watercourses like the Tana, Turkwell, Ewaso Nyiro and Sabaki, other forests like Marsabit and Maralal, and swamps and hilly areas. As elsewhere, conflicts exist between the production sectors. Agriculturalists prefer such areas because the soils are better and water supplies are sufficient for cropping; pastoralists use them as fall-back areas for dry-season grazing; likewise, wildlife converges on them in the dry season. The areas are also preferred for settlement because of water availability and a cooler micro-climate. In broader terms, the conflicts reflect the lack of a co-ordinated national land use policy for the ASAL. Each sector views the key production areas as optimal for their particular activity. What results is intense competition for the same land. Without a mechanism to prioritize competing demands, the resultant land use is not necessarily the most appropriate in ecological terms. For example, many of the key production areas have been gazetted as National Parks or Forest Reserves, which has marginalized the weakest sector, namely, pastoralism.

Conflicts in Wildlife Dispersal Corridors

It is estimated that some 65 to 80 per cent of Kenya's wildlife lives outside National Parks. Many animals move out of the parks in the wet season, while tending to concentrate within them in the dry season when they serve as water and range reserves. Such mobility creates conflicts in the ASAL.

The main conflict outside the parks is between free-roaming animals and agriculture. Crops like maize attract baboons and buffalo, and fields are sometimes poached so badly that it is impossible to grow a crop. Crops are also destroyed by trampling of elephants, buffalo and wildebeest. In Namelock, near Amboseli National Park, it is estimated that no more than 50 per cent of the farming potential is realized due to wildlife competition (MRDASW, 1990). Conflicts also arise between wildlife and pastoralists as a result of stock predation by hyenas, leopards, lions and jackals. There are also problems of disease transmission between wildlife and domestic stock, notably of rinderpest which is transferred by wildebeest and buffalo. Although direct environmental impacts are generally avoided in the above, agricultural activities clearly interfere with wildlife systems.

In some areas, increased fencing of land is blocking movement corridors between wet-season and dry-season wildlife ranges. The disruption of migratory routes reportedly causes species to miss migration or to migrate and fail to return.

Tourism

There is increasing evidence in the ASAL that tourism is detrimental to wildlife. In rangelands around Maasai Mara, tourist camps have proliferated and land use has intensified to the extent that environmental damage occurs. Off-road vehicles cause soil erosion and create noise levels that disturb sensitive animals like cheetahs. Noise also scares birds from their nesting sites, and it is alleged that noise causes some wildlife species to skip mating seasons.

Unless carefully managed for the benefit of local communities, tourism can lead to the displacement and economic marginalization of local populations. Unplanned tourist activities, like hotel and lodge construction, can result in problems of waste disposal and water contamination.

Degradation of Natural Resources in ASAL

Reduced productivity of soil, water, vegetation and wildlife resources occurs as a result of their misuse and overuse, and is the most serious environmental problem in Kenya's ASAL. It constitutes resource degradation and damages production systems. The current extent of such degradation is not well documented, and varies greatly between areas and between resources.

Range Resources

Rangelands experiencing intensified or extended grazing are being exploited at levels that are unsustainable in the long term. This is not a serious problem in most seasonally used rangelands, but, in traditional dry-season grazing areas, increased competition for forage appears to be causing widespread overgrazing. Such overgrazing and associated destruction of ground cover are particularly common around settlements and permanent water courses.

Wood Resources

Overbrowsing occurs in parts of the ASAL, and wood products are often harvested at unsustainable levels. Wood is mainly acquired for fuel, charcoal, furniture and, around towns and other settlements, building purposes. As a result, wood and wood products are in short supply and forests are being destroyed. This deforestation causes soil degradation and depletes water resources.

Rural household consumption of wood is not causing serious deforestation. Rather, it is the urban demand for charcoal and prime timber, like mkoko (*Rhizophora mucronata*), that makes wood harvesting attractive to entrepreneurs. Commercial fuelwood, charcoal and timber production have all had an impact on traditional forest reserves. As timber becomes scarcer, the search for domestic fuelwood becomes an arduous burden that largely falls upon women. Firewood that a few years ago could be collected near dwellings is now carried half a day's walk. In drier areas, like Mandera, Wajir and Garissa, collecting firewood has changed from a task that once took an hour to a chore that takes a whole day.

De-vegetation is now apparent within a 5 to 15 km radius of many settlements in the ASAL. Felling trees for fuelwood and charcoal, clearing land for subsistence farming, as well as overgrazing and trampling by animals around water points are all creating pockets of desert (Darkoh, 1990).

Soil Resources

Soil erosion and salinization are significant local problems in the ASAL. Soil erosion typically occurs where the vegetation cover has been reduced, while salinization results from inappropriate irrigation. Nutrient depletion is also common. Information on the nature and extent of soil degradation is limited and inconsistent, but locally the problem is serious. Top-soil compaction, sheet erosion and gullying occur in districts like West Pokot, Baringo and Samburu. The worst degradation occurs on sloping land in drier areas where land use has intensified and also around settlements. Increased water and wind erosion is also reported around pastoral settlements and permanent water sources in parts of Marsabit District. Some soils are particularly prone to erosion or compaction, and are slow to recover without intervention, e.g.

mechanical pulverization of the surface, reseeding, fencing, terracing and the like.

Problems of the Key Production Areas

A major problem in the ASAL is the degradation of the relatively small areas of high productivity and potential. These key areas are being misused and overused as a result of increased cultivation, wood harvesting, livestock pressure, and urbanization. The number of farmers such areas can support is much smaller than the number of pastoralists. Pastoralism tends to protect soil and water resources better than cropping, and it also allows wildlife to continue using the land. The move towards cultivation in many key production areas has accelerated rates of land degradation, while population growth has contributed to the collapse of traditional land management. Special efforts are thus needed to manage the key areas in the ASAL.

Problems of Water Supply

The water crisis in the ASAL involves both the scarcity of water itself and the land use practices around available water points. Two main types of water resource exist, namely, ground water from boreholes and surface water from dams, pans and reservoirs.

Water points are often the foci of environmental degradation. An average borehole in the ASAL may yield $2000 \, l \, hr^{-1}$, which will supply 2000 cattle or 8000 sheep and goats. Such concentrations of animals around water sources often devastate the surrounding land. The vegetation is completely overgrazed across a large area, and trails are created by herds converging on the water point. The local soils are compacted, so that infiltration rates are lowered and soil erosion increased. Such desertification is especially striking when seen from the air over the North Eastern Province. Poor livestock management also causes water contamination and the spread of disease.

Impacts due to Alteration of Streamflow

Changing land use and the smoothing effects of reservoirs on floods and dry season discharge have a marked effect on the hydrology of some ASAL rivers. This in turn affects riverine vegetation by changing ground water levels and by reducing over-bank flooding and beneficial silt deposition. Traditional farming along rivers, based on recessional crops and on tree crops that tap ground water, may also be affected.

Large projects give even more cause for concern. The Turkwel Gorge Dam is a major hydroelectric project which was constructed between 1986 and 1991. Concern exists about possible environmental impacts, but no major impact assessment has been undertaken. The Turkwel is the largest river system in Turkana District and its gallery forests are a major resource in the dry region through which it passes. Along its 200 km length, some 20 000 to 40 000 people depend on its riverine forest system to provide fodder for their livestock.

While the area near the dam will benefit from the more controlled water flow, the situation is less clear elsewhere in the basin. Thus, only a very small proportion of the water released from the reservoir, when it has eventually filled, is likely to reach Lake Turkana, 200 km away, on account of the high evaporation and infiltration rates. Already Lake Turkana has receded, largely because of a dam on the Oromo river in Ethiopia which supplies 90 per cent of the lake's water; Ferguson Gulf, where 70 per cent of the tilapia in the lake spawned, has dried up. It has been calculated that in the last ten years the level of the lake has fallen by more than a metre. It is not known at what level it will stabilize once the Turkwel reservoir is full and river flow resumed.

Mining Impacts

Mining for building materials, sand, ballast, gypsum, and lime occurs along the coast of Kenya. Some degree of control of these operations has been established. Thus, new annual mining licences are not issued without the approval of the District Environmental Officer, who attempts to ensure that restoration measures, such as backfilling, grading and tree planting, are undertaken once sites are abandoned. However, enforcement of restoration is intermittent and lacks proper direction. In Mombasa, the Portland Cement Company has voluntarily reclaimed wasteland at relatively low cost. Elsewhere in the ASAL, similar reclamation has yet to be undertaken, especially in areas where sand, ballast and fluorspar have been worked.

Development of Traditional Pastoralist Societies

Pastoralist societies in the ASAL have been subject to considerable state control in recent years. The government has sought to integrate their land more fully into Kenya, and an infrastructure of roads, hospitals and schools has been created. Destitute people have been settled on irrigation schemes along major rivers and the shores of larger lakes like Lake Turkana. One effect of these policies has been to restrict the movement of people and animals. This has led to increasing population pressure, and has pushed the pastoralists into marginal areas causing unprecedented environmental damage. Development intervention has also focused resources on agriculture and fisheries rather than livestock, and has concentrated the population into towns and villages (Helland, 1987). This has exacerbated an earlier shift of population towards the heartland of some districts in response to livestock raiding in border areas and towards permanent settlement because of food relief. In Turkana District, for example, such is the concentration of people and livestock that large areas, including some of the best grazing land, are unused, while elsewhere overgrazing and deforestation are occurring, especially in northern Kenya (Sorbo et al., 1988). Many settlements are surrounded by degraded land, which extends as people are obliged to travel further for grazing, fuelwood, and timber for fencing livestock night-enclosures (Lusigi and Glasner, 1984).

In Turkana District, serious environmental degradation has resulted from the concentration of population around irrigation schemes. In such places, farmers are outnumbered by peripheral dwellers, who are seeking work or visiting the area to trade livestock for grain. Deforestation, soil churning and compaction, and water contamination are widespread. Irrigation schemes have also impeded access to rivers and watering points, and have removed browse and grazing resources from pastoral use (Broch-Due and Storas, 1983). These schemes conflict with traditional land use systems and have taken over areas needed for survival by pastoralists in dry years (Darkoh, 1992).

Some researchers and consultants who have worked in the ASAL have been critical of past and present development efforts. They note that, as well as neglecting the pastoral sector, well-meaning government and foreign donor schemes have generally increased human dependence on outsiders, exacerbated human vulnerability to drought, and helped to create a natural environment under increasing pressure (Helland, 1987; Sorbo et al., 1988).

Conclusion

The state of the environment and of land degradation in the Arid and Semi-arid Lands of Kenya is summarized in Table 10.3. The lands are undergoing tremendous changes. Overpopulation and land hunger in the high-potential areas of Kenya are increasing human migration and extending cultivation into marginal areas. As a result, land degradation is a serious problem (National Environmental Secretariat, 1977; Stiles, 1983; Darkoh, 1990). The National Environmental Secretariat of Kenya estimates that, of the total land

Table 10.3 Generalized profile matrix of environmental degradation in Kenya's ASAL.

Factor		*Agro-ecological Zone*			
		IV	*V*	*VI*	*VII*
I	Climate				
	1 Rainfall (annual average in mm)	700–850	550–700	300–500	200–300
	2 Rainfall (50% probability 1st season mm)	250–350	150–200	100–200	–
	3 (50% probability 2nd season mm)	250–350	150–200	50–150	–
	4 Rainfall/evaporation (%)	40–50	25–40	15–25	15
II	Population				
	1 Permanent immigration	high	low	negligible	negligible
	2 Seasonal migration	very high	moderate	low	very low
	3 Return migration	very high	moderate	low	very low
	4 Growth	very high	high	low	very low
	5 Density	very high	high	medium	low
III	Land Use Trends				
	1 Sedentarization of pastoralists	prominent	prominent	moderate	moderate
	2 Conversion of dry season grazing to cultivation	completed	prominent	prominent	prominent
	3 Land privatization	prominent	increasing	beginning	beginning
	4 Displacement of pastoralists	moderate	moderate	prominent	prominent
	5 Effects from land-use changes in highland areas	high	moderate	low	very low
IV	Conflicts				
	1 Land-use conflicts in key areas	low to moderate	moderate	severe	very severe
	2 Wildlife/agriculture	severe	severe	moderate	moderate
	3 Wildlife/livestock	serious	serious	moderate	moderate
V	Resources				
	1 Water resources	ample	poor	very poor	very poor
	2 Vegetation	dry woodland and bushland	bushland	bushland and shrubland	shrubland
	3 Wildlife and tourism	abundant	abundant	abundant	abundant
	4 Livestock production	moderate	high	very high	very high
	5 Crop production	high	moderate	low	very low
	6 Key production areas	abundant	few	few	very few
VI	Degradation of Natural Resources				
	1 Overuse of range (fodder/ bushes) around water sources and settlements	moderate	moderate	high	very high
	2 Overuse of wood resources	high	high	moderate	moderate
	3 Soil degradation	high	moderate	low	very low
	4 Degradation of key production areas	high	high	high	very high
	5 Water supply problems	moderate	moderate	high	very high

Source: after MRDASW (1990).

area of Kenya of 569 137 km², approximately 483 830 km² are already experiencing desertification and other forms of land degradation (Government of Kenya, 1982; UNEP, 1987). This is equivalent to 85 per cent of the country or 90 per cent of the ASAL. Approximately 110 000 km² are already severely affected, while 53 500 km² show latent to moderate degradation.

Although environmental degradation in Kenya reflects climate and current land use, the fundamental problems are socio-economic and political (Baker, 1981; Darkoh, 1990). The rapid growth of human and animal populations has disrupted traditional land use systems that were adapted to the fragile ecosystems of the ASAL. However, growing awareness exists of the problems of land degradation and desertification, and increased emphasis is being placed on public participation in afforestation and in soil and water conservation. Restrictions have been placed on tree clearance and charcoal-making, and support has been given to 'green belt' tree planting organized by local non-governmental organizations. Attention is also being paid to environmental research, and attempts made to slow the rate of population growth through family planning.

Kenyan government policies for preservation and improvement of the environment assume that prevention is better than cure. Current policies emphasize that environmental considerations must be incorporated at the planning stage of development projects (Government of Kenya, 1981). However, either through lack of political clout or of the necessary institutional framework, monitoring of rural-development projects is often lacking. Institutional responses to developmental and environmental problems have been fragmentary, and there is no overall framework setting out the goals and guidelines for environmental protection and resource management (Kinyanjui and Baker, 1980). In a recent study, which sought to evaluate how far the Government had reduced risk and improved environmental and human conditions in the ASAL, it was concluded that hazards had neither been reduced nor ameliorated (Bernard, 1985). Traditional hazard responses in the ASAL have been weakened for both farmers and pastoralists as a result of government intervention. In recent years, many statements have recognized the environmental crisis in the ASAL, but there is disjunction between political pronouncements and institutional responses (DANIDA, 1989). It is essential that any further initiatives in environmental management of the ASAL include a real expression of political will and commitment that matches the seriousness of the situation.

Acknowledgement

Acknowledgement is made of the courtesy of Kenya's Ministry of Reclamation and Regional Development in supplying material from the Environmental Action Plan for the ASAL. The author is responsible for the opinions expressed.

References

Agatsiva, J.L. and Mwendwa, H. (1982) Land use mapping using remote sensing techniques. Paper presented at 4th International Seminar on the Potential Influences of Remote Sensing Survey on Decision-Making, November, Nairobi.

Anzagi, S.K. and Bernard, F.E. (1977) Population pressure in rural Kenya. *Geoforum*, 8, 63–8.

Baker, R. (1981) *Land Degradation in Kenya: Economic or Social Crisis?* Development Studies Discussion Paper No. 82, School of Development Studies, University of East Anglia, Norwich.

Barber, R.G. (1983) The magnitude and sources of soil erosion in some humid and semi-arid parts of Kenya and the significance of soil loss tolerance values in soil conservation in Kenya. In D.B. Thomas and W.M. Senga (eds), *Soil and Water Conservation in Kenya: Proceedings of the Second National Workshop, Nairobi, 10–16th March 1982*. Nairobi: Nairobi Institute of Development, University of Nairobi, pp. 31–43.

Bernard, F.E. (1985) Planning and environmental risk in Kenya's dry lands. *Geographical Review*, 74, 57–70.

Bernard, F.E. and Thom, D.J. (1981) Population pressure and human capacity in selected locations of Machakos and Kitui Districts. *Journal of Developing Areas*, 15, 381–406.

Broch-Due, V. and Storas, F. (1983) *Fields of the Foe. Factors Constraining Agricultural Output and Farmer's Capacity for Participation*. Bergen: University of Bergen/ NORAD.

Campbell, D.J. (1981) Land use competition at the margins of the rangelands: an issue in development strategies for semi-arid areas. In G. Norcliffe and T. Pinfold (eds), *Planning African Development*. Boulder, CO: Westview Press, pp. 39–61.

Campbell, D.J. (1986) The prospect for desertification in Kajiado District, Kenya. *Geographical Journal*, 152, 44–55.

DANIDA (1989) *Environmental Profile: Kenya*. Copenhagen: Ministry of Foreign Affairs.

Darkoh, M.B.K. (1990) Kenya's environment and environmental management. *Journal of East African Research and Development*, 20, 1–40.

Darkoh, M.B.K. (1992) Irrigation and development in Kenya's ASAL. In M.B.K. Darkoh (ed.), *African River Basins and Dryland Crises*. Uppsala: Uppsala University.

English, J., Tiffen, M. and Mortimore, M. (1994) *Land Resource Management in Machakos District, Kenya, 1930–1990*. Washington DC: The World Bank.

Epp, H., Agatsiva, J., Ochanda, N. and Lantier, D. (1982) Application of remote sensing in earth resources monitoring. In *Proceedings of 4th International Seminar on Potential Influences of Remote Sensing Survey on Decision-Making*, Nairobi, pp. 118–27.

Gichuki, F.N. (1991) *Environmental Change and Dryland Management in Machakos District, Kenya 1930–1990: Conservation Profile*. London: Overseas Development Institute Working Paper No. 56.

Government of Kenya (1981) *Economic Survey 1981*. Kenya: Ministry of Economic Planning and Development.

Government of Kenya (1982) *Environment Management Report*. Nairobi: National Environmental Secretariat.

Hansen, S. (1992) Population and the environment. *African Development Review*, **4**, 118–64.

Helland, J. (1987) *Turkana Briefing Notes, Development Research and Action Programme*. Nairobi: Michelsen Institute.

HMSO (1934) *Report of the Kenya Land Commission*. London: His Majesty's Stationery Office.

IFAD and UNDP (1988) *Arid and Semi-Arid Lands (ASAL) Development Issues and Options*, 2 vols. Nairobi: United Nations Development Programme.

Jaetzold, R. and Schmidt, H. (1982) *Farm Management Handbooks of Kenya*, 4 vols. Nairobi: Ministry of Agriculture.

Kinyanjui, D.N. and Baker, P.R. (1980) *Report on the Institutional Framework for Environmental Management and Resource Use of Kenya*. Nairobi: National Environmental Secretariat.

Lusigi, W. and Glasner, G. (1984) Combating desertification and rehabilitating degraded production systems in northern Kenya: the IPAL Project. *Desertification Control Bulletin*, **10**, 29–36.

Mbuvi, J. (1991) Soil fertility. In M. Mortimore (ed.), *Environmental Change and Dryland Management in Machakos District, Kenya 1930–1990: Environmental Profile*. London: Overseas Development Institute Working Paper No. 53, pp. 29–38.

Mortimore, M. (1992) *Environmental Change and Dryland Management in Machakos District, Kenya 1930–1990: Tree Management*. London: Overseas Development Institute Working Paper No. 63.

Mortimore, M. and Wellard, K. (1992) *Environmental Change and Dryland Management in Machakos District, Kenya 1930–1990: Profile of Technological Change*. London: Overseas Development Institute Working Paper No. 57.

MRDASW (1990) *Draft Environmental Action Plan for Sustainable Development for ASAL*. Nairobi: Ministry of Reclamation and Development of Arid, Semi-Arid Areas and Wastelands.

Mutiso, G.C. (1989) Managing Kenya's arid and semi-arid lands. In A. Kiriro and C. Juma (eds), *Gaining Ground. Institutional Innovations in Land Use Management in Kenya*. Nairobi: ACTS Press, pp. 67–104.

Mutiso, S.K., Mortimore, M. and Tiffen, M. (1991) Rainfall. In M. Mortimore (ed.), *Environmental Change and Dryland Management in Machakos District, Kenya 1930–1990: Environmental Profile*. London: Overseas Development Institute Working Paper No. 53, pp. 14–28.

National Environmental Secretariat (1977) Kenya's experience in combating desertification, Kenya's country position paper. Paper presented at United Nations Conference on Desertification, 29 August–9 September 1977, Nairobi.

Sindiga, I. (1984a) Land population problems in Kijiado and Narok, Kenya. *African Studies Review*, **27**, 23–9.

Sindiga, I. (1984b) Inducing rural development in Kenya's Maasai land. *Journal of East African Research and Development*, **14**, 162–77.

Sorbo, G.M., Skjonsbert, E. and Okumu, J. (1988) *NORAD in Turkana. A Review of the Turkana Rural Development Programme*. Nairobi: NORAD.

Stiles, D.N. (1983) Camel pastoralism and desertification in northern Kenya. *Desertification Control*, **8**, 2–8.

Thom, D.J. and Martin, N.L. (1983) Ecology and production in Baringo–Kerio valley, Kenya. *Geographical Review*, **73**, 15–29.

Tiffen, M. (1991) *Environmental Change and Dryland Management in Machakos District, Kenya 1930–1990: Population Profile*. London: Overseas Development Institute Working Paper No. 54.

Tiffen, M. (1992) *Environmental Change and Dryland Management in Machakos District, Kenya 1930–1990: Farming and Income Systems*. London: Overseas Development Institute Working Paper No. 59.

Tiffen, M., Mortimore, M. and Gichuki, F. (1994) *More People, Less Erosion: Environmental Recovery in Kenya*. Chichester: John Wiley and Sons.

UNEP (1987) *Kenya. National State of the Environment Report*. Nairobi: United Nations Environmental Programme.

Wamalwa, B.N. (1989) Indigenous knowledge and natural resources. In A. Kiriro, and C. Juma (eds), *Gaining Ground. Institutional Innovations in Land Use Management in Kenya*. Nairobi: ACTS Press, pp. 45–66.

Wisner, B. (1977) Man-made famine in eastern Kenya: the interrelationship of environment and development. In P. O'Keefe and B. Wisner (eds), *Land Use and Development*. African Environment Special Report. London: International African Institute, pp. 216–28.

11

The Role of Remote Sensing and GIS in Wasteland Management in India

Ravoori Nagaraja and Naresh Chandra Gautam

India is the seventh largest and second most populous country in the world with unique physical and cultural diversity. Diversity in the physical landscape has resulted in many different types of land use, and increasing pressure on land for agricultural and non-agricultural needs has extended to less favourable environments leading to accelerated soil erosion and excessive land degradation. Large areas have been transformed into wasteland, the product of desertification, soil salinization, waterlogging, soil erosion and industrial exploitation. There is an urgent need to reverse this trend and to restore the productive capacity of wastelands in order to meet the demands of increasing population and other developmental needs.

Several agencies have estimated the total extent of wastelands in India. However, their figures vary considerably, ranging between 30 and 175 million ha, in part because of the lack of a mutually agreed definition of wastelands. Reliable information on the location, character and extent of wastelands on a national scale is essential for launching a programme for wasteland improvement and development. In recent years, it has become possible to combine two technologies, i.e. remote sensing and Geographic Information Systems (GIS), which allows a scientific approach to the problems of wasteland identification, improvement and management.

Wasteland Mapping

Until recently, no attempt had been made to prepare a comprehensive map on any scale showing the distribution of different types of wasteland in India. A number of organizations and agencies commenced the collection and

Table 11.1 Area of wastelands in India as reported by different agencies.

Organization	Million ha	Percentage of land area
National Commission of Agriculture, Ministry of Agriculture and Irrigation, Government of India, New Delhi, 1976	45.0	13.7
Directorate of Economics and Statistics, Government of India, New Delhi, 1978–79	38.4	11.7
Directorate of Economics and Statistics, Government of India, 1985–86	39.9	11.2
Department of Environment, Government of India (Vohra, 1980)	95.0	28.9
Ministry of Agriculture and Cooperation, New Delhi, 1984–85	175.0	53.2
Society for Promotion of Wastelands Development, 1984	93.7	28.5
NWDB, Ministry of Environment and Forest, New Delhi, 1985 (NWDB, 1987)	123.0	37.5
NRSA, Department of Space, Government of India, Hyderabad, 1985 (NRSA, 1985)	53.3	16.2

collation of data on the type and extent of wastelands. The area reported by various government agencies is shown in Table 11.1. The considerable variation in the area underscores the need to prepare a reliable data base using the latest techniques of satellite remote sensing and geomatics.

In 1985, the National Remote Sensing Agency (NRSA) of the Department of Space prepared wasteland maps of all the Indian states and union territories at a 1:1 million scale. About 190 Landsat MSS colour composite images (scale 1:1 million) obtained between 1980 and 1982 were used for the study. Visual interpretation based on such image characteristics as colour, texture, pattern, shape, size, location and association permitted the interpreters to identify and delineate eight types of wasteland. After preliminary interpretation, field checks were carried out in doubtful areas and necessary corrections were incorporated into the final map. The wasteland details were transferred to base maps of each state (scale 1:1 million), and, finally, an area estimation of each wasteland category was made.

The total area of wasteland in 1980–1982 was estimated to be 53.3 million ha or 16.2 per cent of the area of the country. The area of different wasteland categories is shown in Table 11.2. The spatial distribution of wastelands in India, imaged by Landsat MSS is shown in Figure 11.1.

The 1:1 million scale wasteland maps of the states and union territories of India provided only a gross estimate of wasteland area. Due to the small scale and relatively poor resolution of the MSS imagery, wasteland areas of less than 100 ha could not be mapped. Furthermore, it was found that the maps were of little use in reclamation planning at the local level.

In 1985, the then Prime Minister of India set up a National Wastelands Development Board (NWDB) with the objective of rehabilitating 5 million ha of land each year for fuelwood and fodder production through a massive

Table 11.2 Area of wastelands in India by category.

Class	Wasteland category	Area (million ha)	
Cultivable wasteland	Salt-affected land	3.90	
	Gullies or ravines	4.32	
	Waterlogged or marshy land	0.89	
	Undulating upland with or without scrub	10.79	
	Jhum or forest scrub	2.40	
	Sandy areas	10.53	
			32.83
Non-cultivable wasteland	Barren hill ridges or rock outcrops	2.75	
	Snow-covered areas	17.70	
			20.45
	Total		53.28

Source: After NRSA (1985).

programme of seeding and afforestation. This programme required a very reliable data base that provided details on the type, extent, location and ownership of wastelands.

Wasteland Classification and Detailed Mapping

Confronted by varying estimates of the extent of wasteland, including the NRSA figure based on remote sensing, it became evident that the NWDB had to provide precise definitions of the various categories of wasteland. The Technical Task Force established by the NWDB proposed a classification system consisting of thirteen categories of wasteland, as follows:

1 Gullied and/or ravined land
2 Upland areas with or without scrub
3 Waterlogged and marshy land
4 Land affected by salinity/alkalinity
5 Shifting cultivation areas
6 Under-utilized/degraded forest land
7 Degraded pasture/grazing land
8 Degraded land under plantation crops
9 Sandy areas (coastal and desert)
10 Mining/industrial wastelands
11 Barren rocky/stony areas
12 Steeply sloping areas
13 Snow-covered and/or glacial areas

The 1:1 million wasteland maps prepared by NRSA showed that 146 districts had wastelands covering more than 15 per cent of their area. These districts were considered as critical (except for border districts) and selected for detailed mapping in Phases I and II (1986–89). In Phase III (1990–92), districts having 5 to 15 per cent coverage of wasteland were mapped, and in Phase IV (1992–93) the remaining seven districts of Madhya Pradesh were examined. This mapping exercise covers about 62 per cent of the area of India.

Since the workload in the project was enormous, all the State Remote Sensing Centres and some central organizations were involved in the mapping task. In order to maintain a consistent standard and a uniform system, a manual of wasteland mapping procedures was prepared giving details of

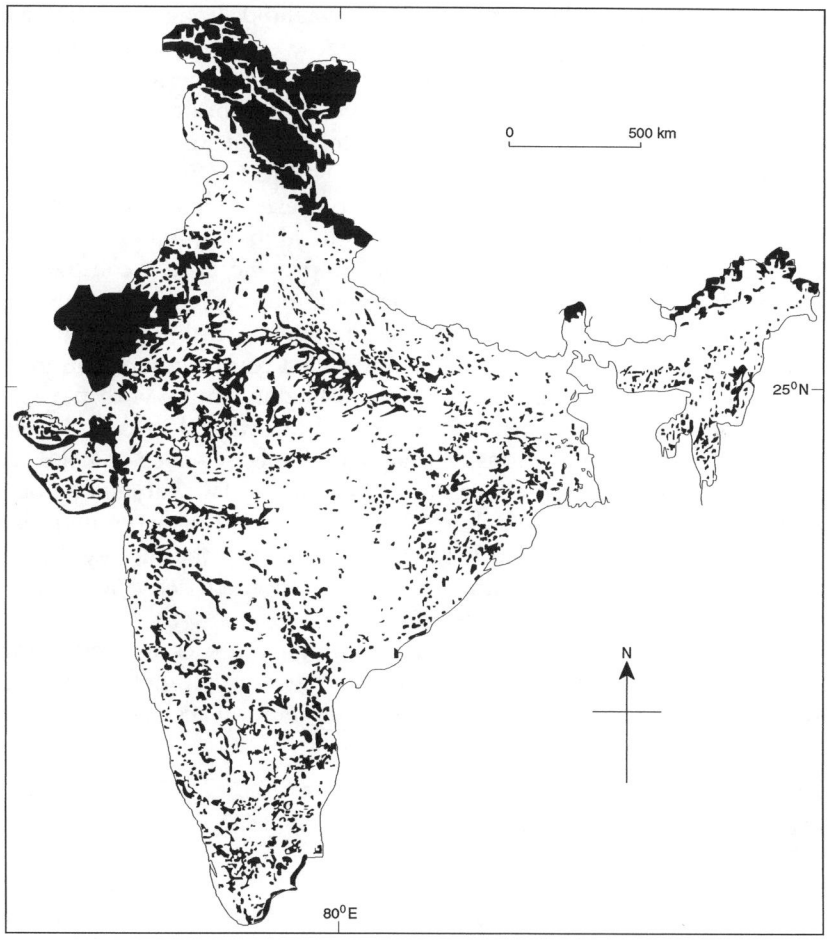

Figure 11.1 Distribution of wastelands in India based on Landsat MSS data.

mapping and final drawing. Training workshops were organized at NRSA to instruct all the scientists involved in the project.

Methodology of Detailed Wasteland Mapping (1:50 000 scale)

The author carried out a pilot study to develop a detailed methodology for the identification and delineation of different types of wasteland using enlarged satellite data. Both Landsat Thematic Mapper (TM) and Indian satellite (LISS-II) data were used for mapping purposes. Visual interpretation of colour composite imagery at 20 × enlargement (1:50000 scale) was undertaken to delineate different types of wasteland. Image characteristics such as colour, texture, pattern, shape and size were used in the process of differentiation. After preliminary interpretation, field checks were carried out in doubtful areas, and corrections were incorporated into the final mapping. An example of the way in which satellite data are incorporated into the mapping process is illustrated in Plate 6. Following final interpretation, the wasteland details were transferred to base maps prepared from topographical maps at 1:50 000 scale using optical instruments.

Village boundaries, as shown on *taluk* maps, and forest compartment boundaries depicted on the forest compartment map were incorporated into the wasteland map using optical transfer methods. The final product is a wasteland map with village boundaries and forest compartment boundaries at 1:50 000 scale. It is possible to associate wasteland units with particular villages using these maps. In addition, possible ownership can be determined by comparison with cadastral maps available at scales of 1:4000 and 1:8000. For the maps being prepared in Phases III and IV, watershed boundaries for drainage basins larger than 5000 ha are being incorporated, using data from the All India Soil and Land Use Survey and the Survey of India. These sheets will help in rehabilitation and development activities based on hydrological units such as river basins.

The areas occupied by each wasteland category are computed using a digital planimeter. The interpretation accuracy using this method has been found to be as high as 85–90 per cent. The interpretation accuracy (IA) can be calculated by applying the following expression:

$$\text{IA} = P \pm \left\{ 1.96 \, \frac{[P \, (100 - P)]}{n} \right\} + \frac{50}{n}$$

where P is the desired proportion of the estimated accuracy expressed as a percentage, n is the minimum number of points (sample size) selected, and 1.96 is a constant (z value assumed at 5 per cent level of significance).

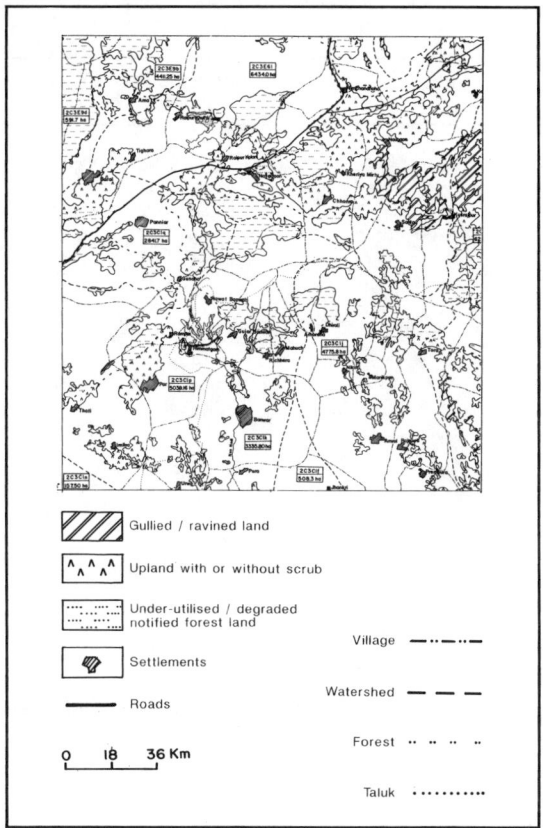

Figure 11.2 Wasteland map of part of Gwalior District, Madhya Pradesh, India.

Results

Under Phases I to IV, about 62 per cent of the national territory has been mapped, with the remaining districts being covered under Phase V. About 36.38 million ha have been designated as wastelands. An example of a wasteland map is shown in Figure 11.2. This is part of Sheet No. WL 54 J/4 (scale 1:50 000) showing the area immediately to the south of Gwalior in Madhya Pradesh. It is one of approximately 4000 wasteland maps completed so far. After completion, the maps are sent to various state users for wasteland reclamation measures. Brief reports for each district provide a general description of the wasteland and its distribution. User interaction workshops have been held in different states to demonstrate the utility of wasteland maps in reclamation activities and to instruct users in making the best use of such materials.

Table 11.3 Percentage frequency of wasteland in each state for the districts mapped in the project.

State	Number of districts mapped	Wasteland as percentage of total area
Andhra Pradesh	19	18.05
Arunachal Pradesh	1	22.43
Assam	2	57.02
Bihar	16	15.02
Goa	2	16.75
Gujarat	13	17.66
Haryana	8	7.81
Himachal Pradesh	3	37.72
Jammu and Kashmir	3	48.32
Karnataka	14	11.93
Kerala	6	4.95
Madhya Pradesh	45	15.81
Maharastra	17	19.55
Manipur	4	37.70
Nagaland	4	54.08
Orissa	13	15.51
Punjab	6	7.48
Rajasthan	20	29.34
Tamilnadu	11	15.65
Uttar Pradesh	27	9.97
West Bengal	3	7.76
Total	237	

Table 11.4 Relative frequency of categories of wasteland for the districts mapped in the project.

Wasteland category	Percentage frequency
Gullied and/or ravined land	5.28
Upland areas with or without scrub	36.40
Waterlogged and marshy land	1.16
Land affected by salinity/alkalinity	3.35
Shifting cultivation areas	4.89
Under-utilized/degraded forest land	28.20
Degraded pasture/grazing land	3.57
Degraded land under plantation crops	1.32
Sandy areas (coastal and desert)	4.12
Mining/industrial wastelands	2.15
Barren rocky/stony areas	7.15
Steeply sloping areas	1.15
Snow-covered and/or glacial areas	1.26
Total	100.00

The distribution of wasteland and the percentage frequency in each state are shown in Table 11.3. The relative frequency of the different categories of wasteland for the 237 districts is shown in Table 11.4. The very high percentage of wasteland in Jammu and Kashmir (48.32 per cent) and in Himachal Pradesh (37.72 per cent) results from the large areas of snow and degraded forest land. In Nagaland (54.08 per cent), Assam (57.02 per cent) and Manipur (37.70 per cent), the wastelands are the product of shifting cultivation, and in Rajasthan (29.34 per cent) they have developed on dune sands. When examined by category, it is found that the upland wastelands (category 2) occupy the largest area, followed by under-utilized forest land (category 6).

GIS in Wasteland Management

Geographic Information Systems (GIS) technology has assumed great importance in the last decade. It is a mapping system with a capability of recording, storing, processing, manipulating and retrieving data. GIS comprises a dedicated computer software for handling geographically referenced spatial data and corresponding attribute information. It has the capability of digitizing maps, of overlaying spatial data, and of displaying information needed in decision-making. Geographic and attribute information can be merged and manipulated in the GIS, which can function as a data base management system. Its graphic display capability allows information retrieval in map form. Input maps are digitized and the GIS software sorts the data in either raster (cell-based) or vector-based systems. There are advantages and disadvantages to both systems. Software developments in the near future promise easy transfer from one to the other.

It is clear that wasteland management needs to be based on an integrated approach, where all relevant factors are considered, including the condition of the wasteland itself, the needs of local people, the availability of development inputs, and the technology for development. The management procedures for different categories of wasteland will vary, depending on the type of degradation, the environmental conditions, and the needs of local people.

The National Wasteland Development Board has approached nine leading organizations in India to undertake pilot studies with different kinds of GIS for different types of wasteland development. The NRSA, using ARC/INFO GIS, has selected Gudekote Mandal in the Bellary district of Karnataka, southern India as a study area. A computerized data base of natural resources and socio-economic information has been generated and procedures based on spatial and non-spatial modelling have been developed for wasteland development planning.

Demonstration Study: Gudekote Mandal

Gudekote Mandal consists of five villages and twenty-one hamlets, covering an area of 18 241 ha. The primary elements of the spatial data base consist of maps showing land use and land cover, soils, ground water potential, surface water bodies and wasteland. The maps are generated using remotely sensed data at 1:50 000 scale. Administrative details, the transport network, and contour information are derived from the Survey of India topographic maps. Secondary elements, including land ownership and also proximity to settlements and roads, are incorporated into the data base.

Digital terrain analysis of Gudekote Mandal was undertaken to generate 3D perspective models of the landscape from spot height and contour information using the TIN module of ARC/INFO. IRS LISS-II data for the area have been rectified using GCPs and registered to the digital elevation model with an RMS error of 1.5 pixels. An example of the merged data sets in

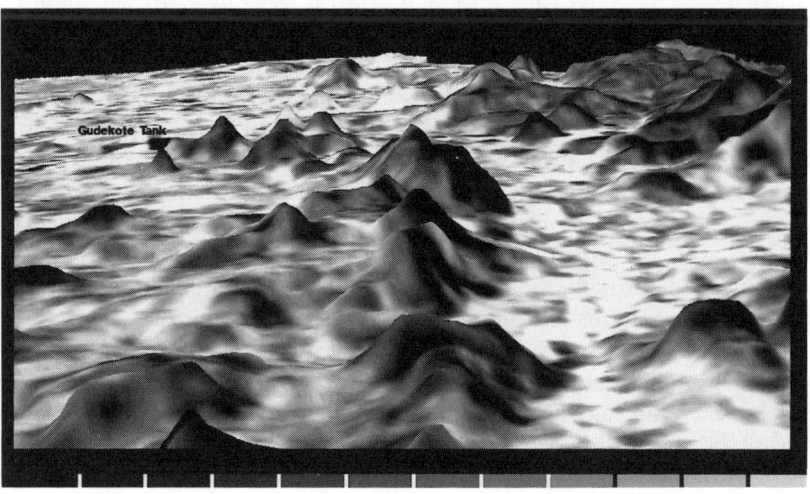

Figure 11.3 Perspective 3D view of Gudekote Mandal, Bellary District, Karnataka, India. View from SSE. Elevation range 520–865 m (VE × 4).

a 3D perspective view is shown in Figure 11.3. Perspective models such as this provide a synoptic view of the terrain and will allow planners and the district's administrator to visualize development proposals, soil-conservation measures and the like. Digital terrain models permit automated landform delineation, mapping of ridges and stream lines, and rapid determination of the boundary characteristics of catchments and watersheds.

The capability of different parts of the study area for different types of wasteland development was assessed by integrating various thematic data sets, including those for soil, groundwater potential and slope conditions. The parameters considered most relevant in the analysis of wasteland are soil depth, erosion status, water potential, slope, distance to roads, distance to water bodies, and the way of life of the local people. The GIS software routines allow the merging of files of various attributes and the identification of Composite Land Development Units (CLDUs). Definition of CLDUs has

Table 11.5 Categories of degraded forest wasteland and proposed land use in Gudekote Mandal.

Class	Area (ha)	Physiography	Slope	Soil	Erosion	Groundwater potential	Proposed land use
1	1838	Residual hills with sheet-rock and boulders	> 15%	Nil	Nil	Nil	Quarrying with due environmental protection
2	1485	Residual hills with sparse vegetation	< 5%	Shallow, well-drained reddish-brown gravelly loamy sand to sandy loam	Moderate	Poor to moderate	Afforestation, e.g. teak, Casuarina, Anogeissus, Terminalia spp.
3	1187	Residual hills with sparse vegetation	5–15%	Very shallow to shallow well-drained gravelly loamy sand	Severe	Poor	Protection of existing species and afforestation with drought-resistant trees, e.g. Prosopis, Acacia, Dalbergia, Eucalyptus, Albizia, Emblica spp.
4	1042	Residual hills with sparse vegetation	> 15%	Very shallow well-drained soil with > 70% rock outcrop and boulders	Very severe	Poor	Soil conservation measures and afforestation with drought-resistant species, e.g. Rubinia, Pseudoacacia, Agave, Euphorbia spp.
5	51	Valley fills	1–5%	Imperfectly drained saline/saline alkaline sandy loam to loamy sand	Nil to slight	Moderate, with moderately saline groundwater	Plantation of salt-resistant trees, e.g. Prosopis, Acacia, Ziziphus, Casuarina spp.
6	255	Moderately deep pediments	0–5%	Moderately deep, well-drained yellowish red loamy sand to gravelly sandy clay loam	Slight to moderate	Moderate	Plantation of economic timber species
7	2956	Shallow pediment	< 5%	Shallow, well-drained dark yellowish brown to dark brown loamy sand to gravelly sandy loam	Moderate to severe	Moderate	Fodder and grassland development
8	300	Shallow pediment	> 5%	Very shallow to shallow, well-drained dark brown loamy sand	Severe	Poor to moderate	Pasture development
9	500	Residual hills with sheet-rock and boulders	> 15%	Nil	Nil	–	Quarrying with environmental protection
10	362	Residual hills with sparse vegetation	< 5%	Shallow, well-drained reddish brown gravelly loamy sand to sandy loam	Moderate	Poor to moderate	Agricultural/horticultural plantations, e.g. cashew, Casuarina, Terminalia catappa, mango, jackfruit
11	329	Residual hills with sparse vegetation	5–15%	Very shallow to shallow, well drained gravelly loamy sand	Severe	Poor	Social forestry and energy plantations, e.g. Anogeissus, Acacia, Prosopis, Ziziphus spp.
12	508	Residual hills with sparse vegetation	> 15%	Very shallow, well-drained soil with > 70% rock outcrop and boulders	Very severe	Poor	Soil conservation measures and afforestation with drought-resistant species, e.g. Agave, Euphorbia spp.
13	32	Valley fills	1–5%	Imperfectly drained, saline/saline alkaline sandy loam to loamy sand	Nil to slight	Moderate, with moderately saline groundwater	Salt-resistant crops – groundwater can be mixed with canal water for irrigation reclamation of salt-affected soils
14	60	Moderately deep pediment	0–5%	Moderately deep, well-drained yellowish red loamy sand to gravelly sandy loam	Slight to moderate	Moderate	Agriculture and horticulture

been achieved by discussion with various resource scientists and local officials. Limiting factors were also taken into account in this process. All wasteland areas are categorized into fourteen CLDU classes, including six classes of under-utilized/degraded forest land and eight classes of upland with or without shrub. An example of the categorization of wasteland and the possibilities for wasteland development in Gudekote Mandal district are shown in Table 11.5.

Conclusions

Wasteland mapping at the 1:50 000 scale has been completed for 237 severely affected districts in nineteen states in India, and in future years the wasteland data base will extend to the whole of India. The maps generated in the project are being widely distributed to various users, including voluntary agencies for reclamation activities. Technologies developed in laboratories of the Council of Scientific and Industrial Research and the Indian Council of Agricultural Research are also being applied in reclamation of wasteland for productive use. The wasteland data base has provided the National Wasteland Development Board and other institutions responsible for financing the reclamation of wastelands with a framework with which to assess priorities.

Wasteland development planning needs specific information if optimal and sustainable development is to be achieved. Satellite remote sensing has clearly demonstrated its usefulness in providing information on the location, extent and type of wasteland. An integrated model for wasteland analysis and planning has been developed and applied by nine organizations using GIS, and the application of this model to the whole of India is imminent.

References

NRSA (1985) *Mapping of Wastelands in India*. Hyderabad: National Remote Sensing Agency.

NWDB (1987) *Description and Classification of Wastelands. Technical Task Group Report*. New Delhi: National Wasteland Development Board.

Vohra, B.B. (1980) *Land and Water Management Problems in India*. New Delhi: Training Division, Department of Personnel and Administrative Reform, Ministry of Home Affairs.

12

Conflicts and Contradictions in Environmental Management in Zimbabwe

Lovemore M. Zinyama

Since the early 1980s, many countries have introduced legislation and other policy guidelines aimed at protecting the environment and ensuring long-term sustainable development. Concurrent with this growing environmental concern, many tropical African countries have sunk deeper into economic crisis. Most have experienced negative economic growth, rising unemployment and poverty, food shortage and hunger, falling export commodity prices, crippling debt burdens, and decreasing inflows of both capital investment and development assistance (World Bank, 1989; United Nations, 1991). Because of the crisis, African governments have been compelled by the International Monetary Fund and the World Bank to implement economic structural adjustment programmes that, so far, give little hope of alleviating the plight of the poor.

In brief, tropical Africa is experiencing a vicious cycle of increasing poverty and environmental degradation that threatens the continental ecosystem (Moyo, 1991; Gore *et al.*, 1992). Poverty means a lack of alternative sources of livelihood, making people even more dependent on the land for their survival. This puts pressure on resources, leading to land degradation, which in turn exacerbates their poverty. Thus, poverty is both a cause and a consequence of land degradation. The problems of poverty alleviation and of environment in tropical Africa represent, not two distinct crises, but one complex crisis with many interrelated facets, which must be tackled simultaneously if the cycle is to be broken.

Environmental degradation as well as policies for environmental management occur within the context of given national social, political, legal and institutional milieux that differ both between countries and within a single country over time. This chapter examines these issues with respect to

Zimbabwe. After outlining the major environmental problems in the country, there follows an examination of existing legislation and policies for environmental management. The evolution of environmental impact assessment as a policy instrument is also discussed. It will be shown that the goals of environmental management and sustainable development often conflict with political and social demands for economic growth. Equally significant, the formulation and implementation of policy have often been shaped by the nature of the power relationships and conflicting interests of different social groups within Zimbabwean society.

Environmental Problems in Zimbabwe

The past century has shown that Zimbabwe's physical environment is highly fragile and sensitive to mismanagement. Degradation of the country's natural resources is particularly a product of land alienation by white settlers, which resulted in the forced relocation of the black population during the colonial period (1890–1980) into ecologically marginal reserves, now known as communal farming areas (Figure 12.1). Today, those parts of the country

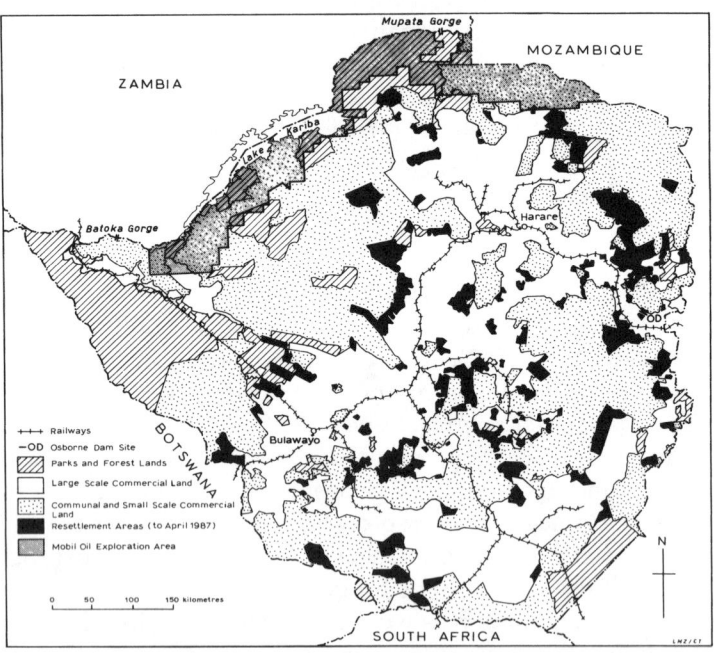

Figure 12.1 Land distribution in Zimbabwe by major land categories, with selected major development projects.

that were reserved for white settlement are sparsely populated, generally well managed and have high standards of environmental protection. Resource conservation was well supported by the colonial government, which, among other things, encouraged the establishment of a strong grassroots network of Intensive Conservation Committees in these areas from the mid-1940s (Whitlow, 1988). Substantial under-utilization of land also occurs within these commercial farming areas (Weiner *et al.*, 1985).

Table 12.1 Distribution of population by major land categories in Zimbabwe in 1982.

Land category	Land area (ha)		Population	
	Total ('000)	%	Total ('000)	%
Communal land	163.9	41.9	4175.2	55.3
Small-scale commercial land	13.2	3.4	166.9	2.2
Large-scale commercial land	158.3	40.5	1253.2	16.6
Other land	55.3	14.2	1950.8	25.9
Total	390.7	100.0	7546.1	100.0

Notes: 1982 census data are used since the final results of the 1992 census are as yet unavailable. The land area data are prior to any major land transfer under the resettlement programme. 'Other land' includes urban land, state forests, national parks, etc.

Areas that were set aside for black peasant farmers comprise 42 per cent of the country (Table 12.1). Today, such areas present the most intractable problems of environmental degradation, being characterized by widespread soil erosion and siltation of rivers, and by vegetation clearance to meet the demands for settlement, cultivation and fuelwood of a rapidly expanding population with a current annual growth rate of 3.1 per cent. The communal areas support some 55 per cent of the 10.4 million people in the country (1992 census). In addition, many urban migrant workers retain their rights to rural land, which is cultivated by their wives and children to supplement inadequate family incomes from wage employment.

The soils in most communal areas, derived from granitic rocks, are sandy, inherently infertile and highly susceptible to erosion when cultivated or overgrazed. Without adequate cash to purchase chemical fertilizers, crop yields quickly decline and the threat of food shortage forces rural dwellers to overexploit their land in a struggle to attain household food security. Increasing population pressure and land shortage over the past century have driven many peasant farmers to clear marginal land on steep slopes and along watercourses, thereby compounding soil erosion and siltation of rivers and dams (Whitlow and Zinyama, 1988; Zinyama, 1988). The prevention of further environmental damage and the rehabilitation of severely degraded lands within the communal farming sector present the greatest environmental challenge for Zimbabwe.

National parks and forest reserves cover 12.7 per cent of the land area. Their primary function is to preserve biodiversity. Most national parks are located in marginal areas with fragile environments that are susceptible to degradation if opened for human settlement and cultivation. Although localized incidents of encroachment by cattle from adjacent communal farming areas may occur occasionally, the major threat in the national parks has emerged in recent years from internationally organized gangs poaching large animal species, notably elephant and rhinoceros.

Contradictions in Environmental Management Legislation

For almost a century, Zimbabwe has had legislation intended to limit wasteful and destructive exploitation of natural resources and to control environmental degradation. The first resource regulations were promulgated in 1913. These were the Water Ordinance and the Herbage Preservation Ordinance, which sought to control stream-bank cultivation, indiscriminate tree felling and bush fires, all of which remain major concerns (Harvey, 1991). The abortive attempt by the colonial administration in the 1950s to change the land tenure in the communal areas to individual ownership under the Native Land Husbandry Act (1951) was motivated by a belief that common ownership of resources was incompatible with good farming practice while the lack of accountability contributed to environmental degradation. At present, a plethora of laws (at least 18 principal Acts of Parliament), plus associated subsidiary regulations, governs environmental and resource issues such as water, soil erosion, stream-bank cultivation, wetland or *dambo* use, mining activities, industrial waste disposal and land use planning.

One of the earliest environmental protection laws was the Water Act (1927) to control stream-bank and *dambo* cultivation. The Act defines the various uses of water and stipulates how these are to be regulated among differing and competing demands. Thus, whereas official permission is not usually required for primary use of public water, i.e. domestic use and watering of livestock, permission is needed for other uses, such as irrigation and industrial uses. The legislation is intended to guard against excessive and uncontrolled abstraction of water from both surface and underground sources and to protect the water rights of other users within a river basin. It also prohibits the disposal of polluting organic or inorganic waste.

The need for an advisory body to the government was recognized early, and led to the establishment of the Natural Resources Board under the Natural Resources Act (1941). The function of the board remains to exercise general supervision over natural resources, to stimulate the dissemination of information on the conservation and management of resources, and to make recommendations on legislation or other measures for the proper management of resources. According to Gore *et al.* (1992, p. 13), the board has generally 'fulfilled its role as a viable and vociferous supporter of grassroots conservation efforts. However, ... its overall impact is subject to the politics of the day.'

The Forest Act (1949) led to the establishment of a Forestry Commission which is charged with the management of state and other forests, the regulation of timber exploitation therein, the conservation of timber resources, and land afforestation. Its companion legislation, the Communal Land Forest Produce Act (1987), specifically regulates the use and management of forest resources within the communal areas. Other laws with a direct bearing on environmental management and protection include the Atmospheric Pollution Prevention Act (1971), the Hazardous Substances and Articles Act (1977), the Parks and Wild Life Act (1975), the Regional, Town and Country Planning Act (1945), the Urban Councils Act (1973), the Rural District Councils Act (1988) and the Communal Land Act (1982).

The most significant legislation for the environment is the Mines and Minerals Act (1961). A person wishing to prospect for, or to exploit, minerals needs a licence issued by a competent authority such as a local Mining Commissioner or the Mining Board. Environmental concerns are not directly addressed when issuing a prospecting or mining licence. The major consideration is that the operation is in the national interest, such as the potential export earnings or new employment opportunities from the project. Furthermore, there are provisions in the Act that make it possible to override environmental protection regulations stipulated in other laws. When a mine is abandoned, the miner is required to fill all shafts and excavations in order to ensure the safety of people and animals. However, there are no specific requirements to protect the environment against subsequent degradation of abandoned mine dumps which may contain toxic materials, as in the case of gold mine tailings.

Analysis of current environmental legislation highlights several issues concerning economic and power relations between different sectors and interest groups. Also of concern is the plethora of laws and regulations that are administered by different government ministries and departments, often in an uncoordinated way. This diffusion of responsibilities reflects the fact that the laws were originally formulated with objectives other than environmental protection. This is clearly shown by the supremacy of the mining legislation over other environmental laws. Mining, together with agriculture, is a vital economic sector in Zimbabwe, with gold, asbestos, chromium as well as tobacco being the principal foreign exchange earners.

Similarly, recommendations of the Natural Resources Board, which is the principal advisor to government on environmental matters, are sometimes overridden for political or economic reasons. One example is the recent expansion in panning for alluvial gold. Until 1991 when new legislation was introduced, panning was illegal under the Natural Resources Act, but was permitted under the Mines and Minerals Act. During the severe drought of 1992–93, many rural people resorted to panning and the illegal sale of gold as a strategy for coping with acute food shortages (Campbell *et al.*, 1989). As a result, panning became a political issue. On the one hand, the Natural Resources Board highlighted the environmental consequences of siltation resulting from panning and recommended that the latter be prohibited. Politicians, on the other hand, emphasized the social and economic plight of the panners, as well as the question of safety since at least 34 panners had died during 1991–92 when unsecured mine shafts collapsed on them. As a result, there was a change in the legislation in 1991 to assist the panners by allowing local district councils to issue them with permits to pan along designated sections of river banks. Each permit holder is allowed to work a 50-m stretch of river bank. By mid-1993, some 270 km of river banks had been designated for panning under the supervision of local district councils. However, even this compromise does not adequately address the environmental issue, particularly as many panners continue to operate outside the designated river banks.

Both national laws and subsidiary local authority bye-laws and regulations exist which govern the disposal of dangerous and environmentally damaging industrial wastes. At the same time, the regulations allow periodic exemptions for certain industries, particularly those deemed to be of strategic economic importance. This poses a serious threat to the environment. A further threat arises because the legislation presupposes the use of the 'best practicable technology', which is ultimately determined by cost considerations, rather than the 'best available technology'. According to Maguranyanga and Tshuma (1991), existing legislation on waste disposal is such that a company commonly finds it cheaper to pay a fine for damaging the environment, assuming it is detected and successfully prosecuted, than to implement remedial protection measures.

Environmental Issues in Land Redistribution

A long-standing issue in Zimbabwe has been the inequitable distribution of land between the whites, who constitute less than 5 per cent of the population, and the majority black population. At independence in 1980, some 6000 white commercial farmers controlled about 15.5 million ha of mainly prime agricultural land, with holdings of up to 8000 ha or more. At the same time, blacks had high hopes of regaining at least some of the better agricultural land in order to meet the demand for land and alleviate environmental degradation within the communal areas. By the end of the 1980s, the government had acquired 2.7 million ha of former white-owned commercial farmland as well as transferring a further 0.6 million ha of vacant state-owned land to resettle some 52 300 black families. Much of the land was acquired on a willing-seller/willing-buyer basis in accordance with the constitutional agreement reached at the Lancaster House conference (1979) under which property rights were entrenched for ten years.

In 1989, the government announced proposals to amend the constitution and related legislation governing land acquisition, in order to enable it to expedite the resettlement of an additional 110 000 black peasant families. The new land policy proposes the transfer of a further 5.0 million ha of commercial farmland for resettlement, plus 1.2 million ha for urban expansion and other public uses. The size of the commercial farming sector will ultimately be reduced from 11.2 million ha in the late-1980s to 5.0 million ha, primarily for the production of economically strategic crops like tobacco, wheat and horticultural products.

In the ensuing debate following publication of the new land policy, several arguments were raised against land redistribution on the scale envisaged. Commercial farmers, through the powerful Commercial Farmers Union (CFU), argued that it would have severe consequences for the economy, particularly through decreased food production for domestic consumers, loss of export earnings, decreased employment for farm workers, and damage to industries dependent on the agricultural sector for markets or raw materials. Of particular relevance here is the argument of the CFU, supported by some researchers (e.g. Whitlow and Campbell, 1989), that such large-scale expropriation would merely extend land degradation onto land that had hitherto been farmed under sound conservation and environmental management. The CFU noted that the dense woodlands previously found on land redistributed in the 1980s had rapidly disappeared as the new settlers cleared the land for cultivation, and in some cases simply sold the timber to urban fuelwood merchants. The CFU argued that the government shared the blame for such degradation because it had selected settlers who had no prior farming experience or training. It contended that inadequate provision of shelter and infrastructure, combined with land use systems that were ecologically inappropriate, had exacerbated the situation. Failure to provide security of tenure on the resettlement schemes and lack of accountability for natural resources had also encouraged their wanton destruction.

Some of the concerns raised by the CFU have been accepted by the government, which has since modified its selection criteria for settlers. In the current land redistribution phase, preference is being given to experienced and trained farmers. Even the Zimbabwe Farmers' Union, which represents the more successful communal and black small-scale commercial farmers, fully supports the revised selection criteria. After all, it is this class of communal farmers who are more likely to be chosen for resettlement under the revised criteria. Even so, the new phase of the resettlement programme is likely to compound the environmental problems of the communal areas and increase rural socio-economic disparities. If the poorest farmers are excluded from resettlement, they will thus be condemned in perpetuity to the already degraded communal areas, where it will be very difficult to improve their economic status or to rehabilitate the land.

Farmer versus Technocrat in *Dambo* Cultivation

Both colonial and post-colonial governments in Zimbabwe have been concerned about crop cultivation along river banks and in *dambo* areas. For both peasant farmers and urban low-income families, *dambo* cultivation is important for domestic food production (Bell *et al.*, 1987; Bell and Hotchkiss, 1989; Whitlow, 1990b). Government technocrats have argued that such cultivation not only led to soil erosion and siltation of rivers and dams, but also diminished the water retention capacity of upstream wetlands, thereby adversely affecting riverflow and the water rights of downstream users. Regulations were introduced to control the practice, including since 1952 a ban on all cultivation within 30 m of a stream and on wetlands. As a result, over the past four decades, *dambo* cultivation has diminished. Some 263 000 ha of *dambo* land exist in the communal areas, but only an estimated 15 to 20 000 ha are currently cultivated (Bell and Hotchkiss, 1989).

For their part, communal-area farmers have always considered *dambos* an important resource with multiple uses. The soils retain water well into the dry season, allowing farmers to grow early season grain staples, such as rice and maize, which are ready for consumption in December to February, some three months before the main dryland crops. The water drawn from shallow wells in *dambos* is used for domestic purposes and livestock watering, as well as for irrigating vegetable gardens. Thus, *dambo* cultivation not only ensures household food security, particularly during the 'hungry season', but also contributes to the quality, variety and nutritional value of the food supply, with surpluses being sold to meet household cash requirements (Bell *et al.*, 1987; Zinyama, 1988; Bell and Hotchkiss, 1989).

However, contrary to the intention of the 1952 legislation, the prohibition of cultivation and consequent introduction of communal livestock grazing in *dambos* appears to have led to their desiccation and erosion (Bell *et al.*, 1987; Lambert *et al.*, 1990; Whitlow, 1992). Grazing appears to compact *dambo* topsoils, which in turn promotes run-off and erosion and reduces sub-surface water movement. A comparison of land uses between gullied and non-gullied *dambo* catchments in two communal areas east of Harare showed that the former had a lower proportion of wetland garden cultivation (17.9 per cent) than the latter (36.4 per cent), contrary to the assumption that cultivation increases erosion. Gullied *dambo* catchments were found to have more land devoted to grazing (82.1 per cent) than non-gullied ones (63.6 per cent), suggesting that grazing contributes to wetland erosion (Whitlow, 1992).

Another study in a communal area south of Harare failed to detect any connection between gully advance and *dambo* cultivation, with the highest rate of gully advance being recorded in a *dambo* with the least cultivation (Bell *et al.*, 1987; Lambert *et al.*, 1990). According to Bell and Roberts (1990, p. 136), the prohibition of *dambo* cultivation was an easy way out for government technocrats who 'were largely blind to the ability of African farmers to cope with seasonal variations in water supply through technologically simple but environmentally effective water management'. Fortunately, other government and non-governmental agencies concerned with improving the nutritional and economic status of peasant farming households are now beginning to encourage the regulated use of *dambos* for garden cultivation.

Recent Changes in Economic and Environmental Policy

Government commitment to the conservation of natural resources is expressed in the National Conservation Strategy published in 1987 (Government of Zimbabwe, 1987). The policy statement has been described as a national recognition of the worsening problem of environmental degradation, an admission of the mistakes and failures of the past, and a statement of the resolve to redress the situation (IUCN, 1988). The main goal of the strategy is to integrate sustainable resource use with every aspect of social and economic development and to rehabilitate degraded resources, particularly in the communal areas.

The year 1990 saw a major change in the government's economic policy, from the socialism of the 1980s towards market-driven capitalism. Much of the 1980s had been characterized by economic stagnation and rising unemployment. It is estimated that during the decade to 1990 the population of working age grew at a mean annual rate of 3.9 per cent, while formal sector employment grew at an annual rate of 1.6 per cent (Government of Zimbabwe, 1991). In 1990, the government commenced its structural adjustment programme aimed at revitalizing the economy and attracting both domestic and foreign capital investment to create more employment opportunities. An integral part of the programme was a reduction in government expenditure, with drastic cuts in social services, notably education and health. Another casualty was the environmental management and protection programmes of the Ministry of Environment and Tourism. The latter is most dramatically illustrated in the curtailment of the anti-poaching operations of the Department of National Parks and Wild Life Management and the retrenchment of some 250 game scouts who previously patrolled the parks.

As the government seeks to attract investment, the issue of environmental safeguards against unscrupulous investors bent on profit maximization also arises. While many laws already exist for the conservation and management of the country's environment and resources, no specific legal requirement yet exists for environmental impact assessment (EIA) of large or potentially damaging development projects. This does not mean that EIAs have not been carried out in Zimbabwe. Indeed, they have been undertaken on an informal basis since the early 1980s, with the initial impetus coming from local non-governmental conservation groups. Lately, both private investors from countries where EIAs are now institutionalized as well as international funding agencies have commissioned impact studies or made them a condition of funding large development projects.

In this respect, the construction of the Kariba dam across the Zambezi river in the late 1950s was preceded by extensive discussion relating to its optimal location, the possible effects of the dam and reservoir on seismic activity along the Zambezi valley, the micro-climatic effects of the reservoir, and the relocation of people and animals from land due to be flooded. However, the first formal EIA in Zimbabwe was only carried out in 1982 to report on two alternative sites for a proposed hydroelectric project on the Zambezi river (Nyamapfene, 1991; Du Toit, 1993; Mubvami, 1993). Non-governmental conservation groups were concerned that the government might opt for a site downstream of the Kariba dam at Mupata gorge, rather than the ecologically less valuable site at Batoka gorge, upstream of the lake.

By mid-1993, some ten EIAs of variable detail and methodological complexity had been undertaken in Zimbabwe. They cover dam construction for hydroelectric and irrigation development, mining development, oil exploration, manufacturing, and road and pipeline construction. It is indicative of the current absence of a legal obligation that the EIA for oil exploration in the Zambezi valley by Mobil Oil was carried out in 1990 only after pressure from conservation groups forced the Ministry of Environment and Tourism to take action when project planning had already reached an advanced stage (Whitlow, 1990a). The project had raised public concern since nearly half the exploration zone lay within protected wildlife areas. In the case of the Osborne dam, the EIA was undertaken in 1989 after initial site works had already started and some families had been relocated (Mubvami, 1993). In this case, work continued in spite of serious misgivings in the EIA about the risk of soil erosion in the catchment area, much of which lies within communal areas, and thus about the lifespan of the dam.

The government has now accepted the principle of environmental impact assessment before any major development project is implemented. In 1991, the Ministry of Environment and Tourism started a review of existing environmental legislation in order to consolidate it, and to strengthen its own competence and expertise in environmental socio-economic analysis. Its aim is to develop, over a five-year period, interim policy guidelines to be followed by formal legislation that will make EIA obligatory, with associated procedures for EIA reporting and review and for impact monitoring.

Conclusion

Tropical environments are inherently fragile and their present state of degradation is largely attributable to human mismanagement. Although sustainable development has become the goal of both national governments and the international community, inherent conflicts exist for developing countries that are simultaneously striving to achieve rapid economic growth and employment for their people. An understanding of the causes of environmental degradation in developing countries such as Zimbabwe requires analysis of these conflicting objectives. In particular, it is necessary to understand the underlying causes of poverty in both rural and urban areas, since it is poverty that drives the poor to overexploit, and hence degrade, the resource base on which they depend for their survival.

Current international economic conditions which, in the present African crisis, have been pressing governments to adopt policies that may in fact accentuate poverty must be revised. Equally important is the resolution of internal conflicts between different interest groups in society and the equalization of access to national resources by all. It remains to be seen whether Zimbabwe can achieve appropriate legislation and institutional arrangements for sustainable environmental protection, without at the same time thwarting the goals of economic growth and employment creation for those whose lives are in urgent need of improvement.

References

Bell, M., Faulkner, R., Hotchkiss, P., Lambert, R., Roberts, N. and Windram, A. (1987) *The Use of Dambos in Rural Development, with Reference to Zimbabwe. Final Report of ODA Project R3869*. London: Overseas Development Administration.

Bell, M. and Hotchkiss, P. (1989) Political interventions in environmental resource use: dambos in Zimbabwe. *Land Use Policy*, **6**, 313–23.

Bell, M. and Roberts, N. (1990) The politics and culture of dambo irrigation in Zimbabwe. In D. Cosgrove and G. Petts (eds), *Water, Engineering and Landscape*. London: Belhaven.

Campbell, D.J., Zinyama, L.M. and Matiza, T. (1989) Strategies for coping with food deficits in rural Zimbabwe. *Geographical Journal of Zimbabwe*, **20**, 15–41.

Du Toit, R. (1993) *Environmental Impact Assessment of the Batoka Gorge Hydro-electric Scheme*. Paper presented at Seminar and Workshop on Environmental Assessment and Review in Zimbabwe, March, Harare.

Gore, C., Katerere, Y. and Moyo, S. (eds) (1992) *The Case for Sustainable Development in Zimbabwe: Conceptual Problems, Conflicts and Contradictions*. Report prepared for UN Conference on Environment and Development, ENDA-Zimbabwe and ZERO, Harare.

Government of Zimbabwe (1987) *The National Conservation Strategy: Zimbabwe's Road to Survival*. Harare: Ministry of Natural Resources and Tourism.

Government of Zimbabwe (1991) *Second Five-Year National Development Plan 1991–1995*. Harare.

Harvey, K. (1991) *Environmental Management in Zimbabwe: An Historical Perspective*. Paper presented at Symposium on Environmental Pollution: Key Issues for Mining and Industry in Zimbabwe, October, Harare.

IUCN (1988) *The Nature of Zimbabwe: A Guide to Conservation and Development*. Gland and Harare: International Union for Conservation of Nature and Natural Resources.

Lambert, R.A., Hotchkiss, P.F., Roberts, N., Faulkner, R.D., Bell, M. and Windram, A. (1990) The use of wetlands (dambos) for micro-scale irrigation in Zimbabwe. *Irrigation and Drainage Systems*, **4**, 17–28.

Maguranyanga, J. and Tshuma, L. (1991) *Legal Aspects of Environmental Protection in Zimbabwe: Key Issues for Mining and Industry*. Paper presented at Symposium on Environmental Pollution: Key Issues for Mining and Industry in Zimbabwe, October, Harare.

Moyo, M. (1991) *Environmental Management in Zimbabwe: The Way Forward*. Paper presented at Symposium on Environmental Pollution: Key Issues for Mining and Industry in Zimbabwe, October, Harare.

Mubvami, T. (1993) *Overview of the Current Status of Environmental Assessment in Zimbabwe*. Paper presented at Seminar and Workshop on Environmental Assessment and Review in Zimbabwe, March, Harare.

Nyamapfene, K. (1991) *Environmental Impact Assessment in Zimbabwe: Some Case Studies*. Paper presented at Symposium on Environmental Pollution: Key Issues for Mining and Industry in Zimbabwe, October, Harare.

United Nations (1991) *Economic Crisis in Africa: Report of the UN Secretary-General Prepared for the Session of the Ad-Hoc Committee of the Whole of the UN General Assembly, 3–13 September, 1991*. New York: United Nations.

Weiner, D., Moyo, S., Munslow, B. and O'Keefe, P. (1985) Land use and agricultural productivity in Zimbabwe. *Journal of Modern African Studies*, **23**, 251–85.

Whitlow, R. (1988) Soil conservation history in Zimbabwe. *Journal of Soil and Water Conservation*, **43**, 299–303.

Whitlow, R. (1990a) Mining and its environmental impacts in Zimbabwe. *Geographical Journal of Zimbabwe*, **21**, 50–80.

Whitlow, R. (1990b) Conservation status of wetlands in Zimbabwe: past and present. *GeoJournal*, **20**, 191–202.

Whitlow, R. (1992) Gullying within wetlands in Zimbabwe: an examination of conservation history and spatial patterns. *South African Geographical Journal*, **74**, 54–62.

Whitlow, R. and Campbell, B. (1989) Factors influencing erosion in Zimbabwe: a statistical analysis. *Journal of Environmental Management*, **29**, 17–29.

Whitlow, R. and Zinyama, L. (1988) Up hill and down vale: farming and settlement patterns in Zimunya Communal Land. *Geographical Journal of Zimbabwe*, **19**, 29–45.

World Bank (1989) *Sub-Saharan Africa: From Crisis to Sustainable Growth*. Washington DC: World Bank.

Zinyama, L.M. (1988) Changes in settlement and land use patterns in a subsistence agricultural economy: a Zimbabwe case study, 1956–1984. *Erdkunde*, **42**, 49–59.

PART IV

Degradation in Tropical Wetlands

13

Tropical Wetland Degradation and Strategies for Management

Michael J. Eden

Wetlands comprise 'areas of submerged or water-saturated lands' (Gopal *et al.*, 1982) and they have long been a focus of human activity. This certainly applies in the tropics, where pre-modern exploitation of wetlands for agricultural and other purposes was widespread (Denevan and Turner, 1974; Farrington, 1985; Roosevelt, 1991) and where contemporary wetland development proceeds on a large and expanding scale. Contemporary development includes attempts to exploit wetlands as such, but they are also seen as wastelands to be reclaimed by draining or in-filling so as to make better use of the land (Gopal *et al.*, 1982).

Contemporary human impacts on wetlands are paralleled by a growing awareness of wetland functions and values, but, as in other tropical contexts, systematic scientific appraisal as a basis for land management has been slow in coming. A long-established scientific interest has existed in specific wetland types, e.g. mangrove forests or rice fields, and in particular wetland functions, e.g. as refuges for migratory birds, but it has only been during the 1980s that a broader perspective on wetlands has emerged. This is now leading to investigation of the range of tropical wetlands and assessment of their overall levels of disturbance and degradation.

In this introductory chapter, the nature of tropical wetlands is outlined. Consideration is then given to the human impacts thereon, which range from the limited modification of existing wetlands to the conversion or total loss of wetland areas through artificial drainage or in-filling. Finally, attention is paid to management strategies that offer a more balanced and integrated treatment of wetland degradation.

Wetland Ecosystems

One reason why wetlands have not been treated as an entity in the past is their diversity. Tropical wetlands are mostly Holocene alluvial landscapes, but they display pronounced hydrologic and edaphic variability. They include river floodplains that are prone to severe seasonal inundation, and alluvial overflow plains, commonly tectonic in origin, that suffer less profound seasonal flooding. In addition, there are extensive tracts of estuarine, deltaic and coastal sediments, where tidal rather than seasonal flood regimes occur and where brackish rather than fresh waters prevail. The inherent diversity and variability of these habitats give rise to vegetation that is often successional in character and locally variable in structure and composition (Salo *et al.*, 1986). The vegetation is variously dominated by woody and herbaceous communities, which in the past have often been treated as classes of forest or savanna rather than explicitly designated as wetland.

Nowadays, tropical wetlands are increasingly being seen as a vegetation formation in their own right, although neither their overall extent, nor their level of human disturbance is accurately known (Barbier, 1993). Broad estimates suggest that some 264 million ha of wetland exist in the tropics, occupying 4.8 per cent of the land area (Mitsch and Gosselink, 1993). Extensive wetlands are present in the Southeast Asia/West Pacific region; some 35 million ha of Indonesia, for example, consist of swamp wetland (Donner, 1987). There is also a large extent of wetland in tropical South America, including an estimated 93 million ha in Amazonia (Eden, 1990).

Wetland vegetation is adapted to the temporal variability of its habitats. This has led to the assumption that wetland ecosystems are relatively resilient, their component species having evolved in response to variable or unpredictable environmental conditions (Lowe-McConnell, 1977; Eden, 1990). Such resilience, which contrasts with the apparent fragility of the tropical rain forest (May, 1975), is assumedly an asset in terms of systemic recovery from disturbance, but it does not prevent wetlands from being severely degraded as a result of gross human impacts. Indeed, the natural productivity of wetland systems has commonly attracted human exploitation and settlement, thereby increasing the risk of degradation.

Yet where wetland systems escape being severely degraded by humans, they fulfil natural functions that help to sustain the systems themselves as well as to benefit their human populations. Such functions have been described as 'indirect-use values' (Barbier, 1993), and include, for example, the flood control and storm protection functions of riparian swamp forests and coastal mangroves. Other such functions include sediment and nutrient retention and the maintenance of water quality. It is increasingly recognized that indirect-use values, along with 'non-use' values like the maintenance of wetland biodiversity (Barbier, 1993), need consideration when wetland developments are proposed, since loss of such values implies a degree of wetland degradation.

Wetland Exploitation and Degradation

The natural resources of tropical wetlands have long attracted human populations, and settlement has frequently occurred in or along riparian and coastal zones. The attraction of natural resources has been reinforced by the communication and transport potential of rivers and coastal waters, and, in modern times, these factors have increasingly encouraged the development of urban–industrial activities.

Over the millennia, tropical wetlands have provided abundant fish and game resources and have allowed farmers to exploit a variety of productive habitats. Distinct agricultural systems have evolved, ranging from recessional or post-flood cultivation (Meggers, 1971; Thom and Wells, 1987), through raised-field or ridged-field cultivation (Denevan and Turner, 1974), to wet rice cropping systems (Chang, 1987). The long-term development of these agrosystems has inevitably caused a degree of environmental disturbance and degradation, but, in many areas, the agrosystems have been sustained for extended periods.

In places, traditional wetland agrosystems persist to the present, but they are steadily being superseded by modern systems that require substantial investment in drainage and irrigation works and recurrent inputs of fertilizers and biocides. In areas of wet rice cultivation, yields have increased in recent decades, but there has been additional degradation in the form of soil salinization, eutrophication of water bodies, and the loss of fish resources (Moulton, 1973; Wade, 1980). Given the socio-economic as well as ecological costs involved, a considerable incentive exists to limit this degradation and safeguard the increased crop yields, but this is not always practicable in established rice cultivation areas. Equally, in areas of recent wetland colonization like those associated with the Indonesian Transmigration Programme, land reclamation and cultivation have had mixed results for various environmental and other reasons, and have led to serious land degradation and abandonment of settlement (Donner, 1987; Hurst, 1990; Rich, 1994). A recurrent environmental problem in coastal and estuarine areas in Indonesia and elsewhere is that of acid sulphate soils, which undergo extreme acidification when drained for agricultural purposes (Young, 1980; Donner, 1987).

Elsewhere, other land developments have caused major damage to wetlands. In the vicinity of expanding urban areas, more or less complete conversion or loss of wetland habitat commonly derives from land drainage or in-filling for industrial, commercial or housing purposes. Such loss of wetland is likely to be irreversible. The process is illustrated by Asangwe (Chapter 16) in respect of Lagos, Nigeria, where rapid urban expansion is occurring across extensive coastal wetlands; similar expansion into stands of coastal mangrove forest around Belize City is reported by McShane (Chapter 15). In addition to the direct impacts of urban expansion, indirect damage frequently occurs through the release of sewage, industrial pollutants and other waste products into peri-urban wetlands.

Other forms of development also inflict indirect damage on wetlands, especially on those located in the lower reaches of large catchments (Maltby and Dugan, 1994). Such developments range from hydroelectric dams and other water-management schemes that modify flood regimes and sediment loads downstream to the alluvial gold mining that pollutes river waters with mercury. In addition, the deforestation of free-draining land for agricultural or other purposes modifies fluvial run-off and sediment loads, with implications for downstream wetlands and their inhabitants (Barrow, Chapter 14). In broader terms, rising sea level as a function of global warming constitutes a similar threat to coastal and lower riparian wetlands and their inhabitants (Sattaur, 1990; Mahtab and Karim, 1992).

Wetland Management Strategies

Modern wetland development at times achieves sustainable economic and other benefits that appear to outweigh any associated environmental degradation. On other occasions, development merely produces short-term gains, along with significant and much less justifiable levels of environmental degradation. In recent years, wetland researchers have begun more rigorous economic appraisals of wetland development projects and are providing a more balanced perspective on the costs and benefits involved. Such appraisals are initially concerned with the benefits of a specific development, be it an aquacultural project or a hydroelectric scheme, but they also include any benefits that are foregone as a result of the development; these may be (a) other direct-use values of the wetland, e.g. subsistence fishing or fuelwood collection, (b) the indirect-use value of the wetland, e.g. for flood control, storm protection, or the maintenance of water quality, and (c) the non-use value of the wetland, e.g. for maintenance of biodiversity (Barbier, 1993).

This approach, which is equally applicable in other tropical habitats, albeit as yet scarcely applied there, will expose the true cost of a non-sustainable exploitation of a wetland. Also it raises questions about the net benefit of some productive and apparently sustainable wetland developments. The net benefit of mangrove clearance for specific agricultural purposes in Fiji, for example, is reported to be less than if the mangrove were maintained, when the full range of its benefits is accounted for (McShane, Chapter 15). Similar findings are reported from African floodplain wetlands, such as the Niger inland delta in Mali and the Kafue flats in Zambia. In such areas, existing traditional multiple uses of wetland yield a higher financial return than agricultural conversion projects, mainly because of the capital investments necessary in engineering (Marchand, 1987; Maltby and Dugan, 1994).

This 'total valuation' approach emphasizes the desirability of adopting more conservative management strategies towards tropical wetlands (Turner, 1991; Barbier, 1993). This may imply an increased level of protection of existing natural or semi-natural wetlands or merely greater efforts to maintain the traditional diverse multiple uses and natural functions of wetlands as a means of maximizing their value and limiting their degradation. Given current levels of population density and resource use, explicit wetland protection is not easily achieved in many parts of the tropics, but is increasingly desirable and, in places, beginning to be realized. In Costa Rica, for example, some 10 per cent of existing wetlands have acquired legal protection (Zürcher, 1991). Managing wetlands for both protection and traditional production is also difficult, although, as indicated, an initial assessment of the total economic value of a wetland system, including its indirect-use and non-use values (Barbier, 1993), is a relevant starting point. The need then exists to promote this approach among politicians and planners, so that more balanced wetland management is actually implemented.

In addition, it has to be recognized that local wetland management cannot operate in isolation. Even the most rational and balanced management plan for a specific wetland will falter if the wetland suffers significant external impacts. The extreme case is that of rising sea level, associated with global warming, that may affect coastal and lower riparian wetlands. Of more immediate concern is the human disturbance of upstream terrestrial or aquatic systems that alters the quantity or quality of water supplied to downstream wetlands. Widespread human disturbance of this kind occurs nowadays, and increasingly is affecting the status and dynamics of wetland systems and the productivity and welfare of their human inhabitants. The nature of the external impacts varies. They may be of local origin, involving the flow of urban pollutants to a peri-urban wetland, or of distant origin, reflecting land cover changes, mining activities or hydroelectric projects that affect wetlands far downstream. In either case, major environmental and human impacts can result.

In these circumstances, wetland management needs to be widely integrated. It should be able to operate, as necessary, at national or international level in order to ensure that specific land developments take account of inherent environmental linkages. The concept of watershed management is by no means new to the tropics (Dasmann *et al.*, 1973), but has been neglected of late in so far as recurrent examples exist of development projects that have been narrowly conceived and overlook these linkages. This is particularly so with hydroelectric schemes and other large-scale water-management projects (Barrow, 1988; Cummings, 1990), but also applies to other wetland developments.

The possibility of restoring degraded wetlands exists. Some wetland degradations are effectively irreversible, but many salinized soils and polluted waters can be restored and mangrove forests can be replanted. Yet such restoration is costly and often beyond the resources of the countries concerned, even if they have the political will to proceed. This confirms the desirability in economic, as well as environmental, terms of pursuing effective wetland management in the first place, and thereby avoiding the need for later restoration (Maltby and Dugan, 1994).

Conclusion

Tropical wetland systems are supposedly resilient, but they have long been exploited by humans and are increasingly being degraded by them. Current wetland degradation is primarily due to the direct effects of wetland developments, of which some are sustainable, but it also reflects external developments that indirectly impinge on wetlands. There is a growing awareness of wetland values and functions in the tropics, but the issue of wetland degradation has received much less attention than, say, desertification or deforestation, and as yet an inadequate political commitment exists to developing more balanced and integrated management strategies. Wetland habitats are variable and experience diverse human impacts, making them complicated to manage. Yet they will surely continue to be exploited and developed, and appropriate management strategies need to be formulated and implemented. In this context, the following are required:

1   The adoption of a more balanced perspective on wetland uses and functions, and recognition of the utility of a 'total valuation' approach when considering wetland development.
2   An acceptance of the potential economic, as well as environmental, benefits of setting aside wetlands for various protective functions and/or traditional productive uses.
3   A recognition of the inherent linkages that exist between wetlands and adjoining terrestrial and aquatic systems, and of the value of an integrated perspective as an aid to minimizing wetland degradation.

References

Barbier, E.B. (1993) Sustainable use of wetlands. Valuing tropical wetland benefits: economic methodologies and applications. *Geographical Journal*, **159**, 22–32.

Barrow, C. (1988) The impact of hydroelectric development on the Amazonian environment: with particular reference to the Tucurui project. *Journal of Biogeography*, **15**, 67–78.

Chang, T–T. (1987) The impact of rice on human civilization and population expansion. *Interdisciplinary Science Reviews*, **12**, 63–9.

Cummings, B.J. (1990) *Dam the Rivers, Damn the People. Development and Resistance in Amazonian Brazil*. London: Earthscan Publications.

Dasmann, R.F., Milton, J.P. and Freeman, P.H. (1973) *Ecological Principles for Economic Development*. London: John Wiley and Sons.

Denevan, W.M. and Turner II, B.L. (1974) Forms, functions and associations of raised fields in the Old World tropics. *Journal of Tropical Geography*, **39**, 24–33.

Donner, W. (1987) *Land Use and Environment in Indonesia*. London: C. Hurst and Company.

Eden, M.J. (1990) *Ecology and Land Management in Amazonia*. London: Belhaven.

Farrington, I.S. (ed.) (1985) *Prehistoric Intensive Agriculture in the Tropics*, 2 vols. Oxford: BAR International Series 232.

Gopal, B., Turner, R.E., Wetzel, R.G. and Whigham, D.F. (1982) *Wetlands, Ecology and Management*. Jaipur: National Institute of Ecology and International Scientific Publications.

Hurst, P. (1990) *Rainforest Politics. Ecological Destruction in South-East Asia*. London: Zed Books.

Lowe-McConnell, R.H. (1977) *Ecology of Fishes in Tropical Waters*. London: Edward Arnold.

Mahtab, F.U. and Karim, Z. (1992) Population and agricultural land use: towards a sustainable food production system in Bangladesh. *Ambio*, **21**, 50–5.

Maltby, E. and Dugan, P.J. (1994) Wetland ecosystem protection, management, and restoration: an international perspective. In S.M. Davis and J.C. Ogden (eds), *Everglades. The Ecosystem and its Restoration*. Delray Beach: St Lucie Press, pp. 29–46.

Marchand, M. (1987) The productivity of African floodplains. *International Journal of Environmental Studies*, **29**, 201–11.

May, R.M. (1975) The tropical rainforest. *Nature*, **257**, 737–8.

Meggers, B.J. (1971) *Amazonia. Man and Culture in a Counterfeit Paradise*. Chicago: Aldine Atherton.

Mitsch, W.J. and Gosselink, J.G. (1993) *Wetlands*. New York: Van Nostrand Reinhold.

Moulton, T.P. (1973) More rice and less fish – some problems of the 'Green Revolution'. *Australian Natural History*, **17**, 322–7.

Rich, B. (1994) *Mortgaging the Earth. The World Bank, Environmental Impoverishment and the Crisis of Development*. London: Earthscan Publications.

Roosevelt, A.C. (1991) *Moundbuilders of the Amazon. Geophysical Archaeology on Marajo Island, Brazil*. San Diego: Academic Press.

Salo, J., Kalliola, R., Häkkinen, I., Mäkinen, Y., Niemelä, P., Puhakka, R. and Coley, P.D. (1986) River dynamics and the diversity of Amazon lowland forest. *Nature*, **322**, 254–8.

Sattaur, O. (1990) Guyana's test at high tide. *New Scientist*, **125** (1710), 46–9.

Thom, D.J. and Wells, J.C. (1987) Farming systems in the Niger inland delta, Mali. *Geographical Review*, **77**, 328–42.

Turner, K. (1991) Economics and wetland management. *Ambio*, **20**, 59–63.

Wade, R. (1980) India's changing strategy of irrigation development. In E.W. Coward Jr (ed.), *Irrigation and Agricultural Development in Asia. Perspectives from the Social Sciences*. Ithaca: Cornell University Press, pp. 345–64.

Young, A. (1980) *Tropical Soils and Soil Survey*. Cambridge: Cambridge University Press.

Zürcher, M.H. (1991) The nineties: another 'lost decade' for Latin American wetlands? *IUCN Wetlands Programme Newsletter*, Gland, **4**, 12–13.

14

Environmental Impact of Resource Use on Wetland and Riverine Habitats in Amazonia

Christopher J. Barrow

Although it is an oversimplification, the division of Amazonia into drylands (*terra firme*) and wetlands (*várzea*) contributes to the understanding of the region and the problems of development (Eden, 1990). The Amazonian *terra firme* typically has infertile, phosphate-deficient and aluminium-rich soils. More fertile soils occur locally, but insect pests and weeds as well as difficult and costly road access make sustained crop production a challenge (Goodland and Irwin, 1975; Goodman and Redclift, 1991). Since the 1960s, some 60 to 65 million ha of the *terra firme* (covering about 10 per cent of Amazonia) have been deforested, and calls have been made to divert development to the savannas to the south and northwest, to areas of secondary regrowth, and to wetlands.

It is difficult to assess the extent of wetlands in Amazonia, but they are reported to occupy roughly 90 to 95 million ha (Eden, 1990). They are most extensive along the middle and lower Amazon valley, along larger, turbid or 'white water' tributaries, and in the estuarine zone (Figure 14.1). Amazonian wetlands and riverine environments have long been exploited, albeit until recently at only low levels of intensity (Hall, 1989). The natural resilience of floodplain habitats allows more intensive exploitation, but caution is needed. Amazonian wetlands are an important source of biodiversity, and increased exploitation will destroy valuable crop, timber and aquatic species. In addition, the zone will increasingly be exposed to damaging physical impacts.

Exploitation of Wetland Resources

Black earth soils (*terra preta do indio*) often exist along the margin of the *terra firme* overlooking the *várzea* zone. They are indicative of prolonged pre-

Columbian settlement (Denevan, 1966; Roosevelt, 1989; Grenard and Grenard, 1993). After the conquest, settlers of mixed blood (*caboclos*) established themselves along the *várzea*. They adopted many indigenous practices, living by fishing, extraction of forest products and shifting cultivation (Eden and Andrade, 1988; Nugent, 1993).

Caboclo numbers have remained low and the *várzea* has generally suffered less degradation than the *terra firme*. Most of Brazil's natural rubber (*Hevea* spp.) is extracted from the *várzea*, with the industry currently supporting some 400 000 people. Between the 1930s and 1970s, many *caboclos* adopted jute (*Corchorus capsularis*) or malva (*Sida rhombifolia*) as cash crops in their rotational cropping. This caused little environmental damage and provided a useful income until plastics and artificial fibres undercut the market. Smallholders have made few attempts at other types of development on the *várzea*, not least because of insecure land ownership and the sharecropper/patron

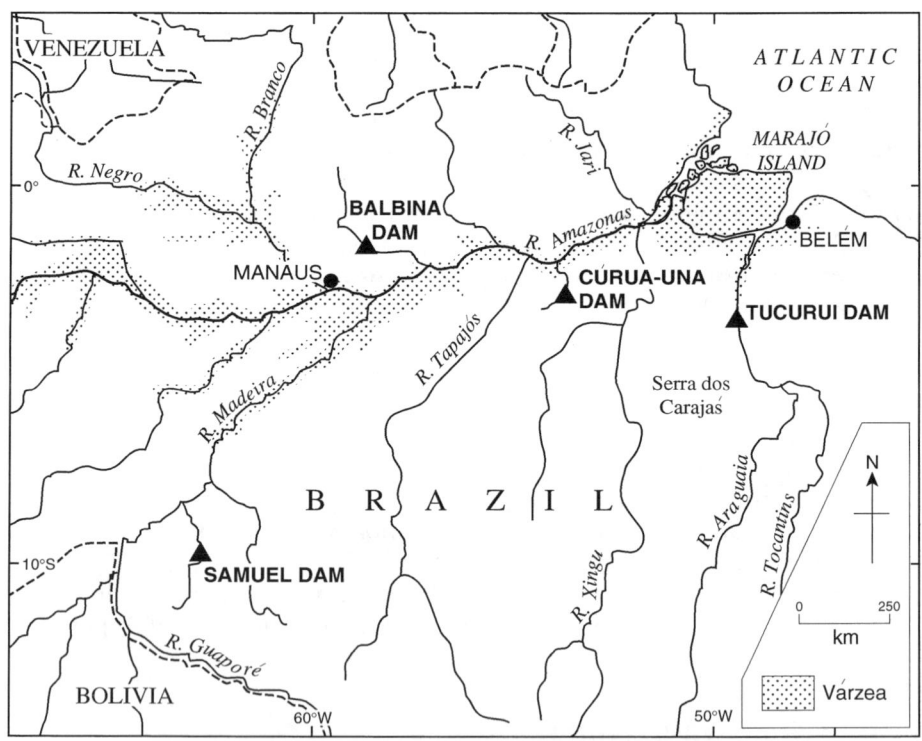

Figure 14.1 Location map of eastern and central Amazonia, showing the *várzea* zone.

relationship that currently controls much of Brazil's extractive and agricultural activities (Gray, 1990).

Agriculture

The development of agriculture on the *várzea* has long been advocated (Lima, 1956; Prance, 1990; Cleary, 1991; Serrão, 1994), but only lately have viable projects begun to attract investment. Large-scale irrigated rice production has been established in the Jari and lower Araguaia valleys (Hall, 1989), but access to land remains a problem. Some of the land belongs to the Brazilian Navy, property titles are often dubious, and widespread confusion exists over many *várzea* land rights. Rice cultivation is also discouraged by volatile grain prices and competitive rice production in adjacent savannas (Madeley, 1993).

Rice yields from the Amazonian *várzea* can be high, and some areas have sufficient silt deposition to preclude the need for chemical fertilizers. With careful drainage and land management, soil degradation can be avoided, herbicide use can be minimized, and sustained cropping of one or two harvests a year is feasible. The *várzea* of the lower Amazon is less favourable, being prone to salinization and in places suffering acid-sulphate toxicity with cultivation.

Extractive Production

Extractive logging has long occurred on the lower Amazon *várzea*. The main species currently exploited are ucuuba (*Virola surinamensis*), andiroba (*Carapa guianensis*), sumauma (*Ceiba pentandra*) and assacu (*Hura crepitans*) (Macedo and Anderson, 1993). Ucuuba is used for plywood, particle-board, paper-pulp, veneer and timber. It is now Brazil's second timber export after mahogany (*Swietenia macrophylla*).

Many Amazonian palms have a commercial potential, notably açaí (*Euterpe* spp.), babaçu (*Orbignya phalerata*), peach palm (*Bactris gasipaes*), patauá (*Jessenia bataua*), jauari (*Astrocaryum flexuosa*) and buriti (*Mauritia flexuosa*) (Barrow, 1990; Kahn and de Granville, 1992). Açaí fruit and palm hearts are readily available throughout Amazonia, and in some areas their collection comprises the main source of livelihood (Anderson and Jardim, 1989; Lopez-Parodi and Freitas, 1990). Of late, açaí production has rapidly increased on the *várzea*, the main markets for the fruit being Belém and Manaus.

Babaçu prefers less swampy wetlands and often invades cleared forest land. It is widely exploited in Maranhão and Goias in Brazil and also in parts of Colombia and Bolivia (Hall, 1989). It is a source of food, oil, charcoal, fibre and thatch, and is a major source of income for an estimated two million people in Brazil. Where the babaçu stands are open, cattle grazing is possible (Alcorn, 1990).

There is increasing interest in 'tolerant forest management' on the *várzea*, i.e. useful forest species are encouraged and understorey species are thinned to improve access (Anderson *et al.*, 1985; Anderson, 1990a). Profitable and sustained management of this kind has recently been described from several *várzea* areas, including Combú island on the Guamá river near Belém (Anderson, 1992). Inundated by seasonal and tidal flooding, Combú island, covering 15 km^2, has a population of some 600 people, who can make a steady living by extracting forest products without much investment of cash or labour. The minimal level of forest management consists of cutting paths to gain access to açaí, rubber, cacao (*Theobroma cacao*) and other useful trees. Supplementary cultivation and fishing are also practised (Anderson, 1992).

Livestock Production

Caboclos traditionally graze cattle on *várzea* meadows during low-water periods. During the flood season they move them to higher ground or on to raised wooden platforms. According to McGrath *et al.* (1993), cattle herds more than doubled between 1974 and 1984, and at least half of *várzea* families now own some cattle, with grazing being most widespread in the lower Amazon. Water buffalo have been established in the area, and improved cross-breeds comprise an increasing proportion of the total cattle stock. Growing stock numbers mean that there is an increasing risk of damage to crops. Large herds are kept near Manaus and on Marajó Island; the latter has extensive wetland grazing, which, in 1978, supported at least 1.43 million cattle.

Impacts on Wetland Ecosystems

Important tree crops originate in the Amazonian wetlands, including cacao, rubber and quinine (*Cinchona officinalis*), and other useful species will surely emerge. Many Amazonian plant species depend on flood dispersal of their seeds or require flood water to grow. The zone also provides breeding, feeding and refuge habitats for fish and other animals. The primary productivity of many Amazonian rivers is poor, and consequently many animals are dependent on flooded forests. Thus, the degradation of these areas will jeopardize the survival of valuable plants and will disrupt fisheries and wildlife. Wetlands also regulate river flows, and wetland degradation may alter flood levels.

In 1981, wetlands yielded about 60 per cent of Amazonian timber (FAO, 1993). Unfortunately, logging controls are poorly enforced in the zone, as elsewhere, and forest stands are being rapidly depleted. Areas around Marajó Island and near the Jari project in particular have been heavily logged.

Wetlands may be drained and cultivated or they can provide water supplies for pump irrigation of the adjacent *terra firme*. There are hazards associated with both types of development. Altered water levels and the receipt of return flows contaminated with sediment or agro-chemicals could have serious impacts on aquatic systems. So far such developments have been limited, but large-scale wetland rice cultivation in particular could expand in the future (Hiraoka, 1993). One of the greatest threats is the intensive use of agro-chemicals which will enter streams and cause damage to wildlife. Incipient problems exist on the lower Tocantins river in eastern Amazonia, where small farmers apply pesticides like aldrin, Dimecron-50 and Mirex, with little apparent advice from extension officers. Amazonia still has large areas of undeveloped wetlands, but the threats to them are manifold.

Exploitation of Riverine Resources

Fisheries

The waters of the Amazon support a rich diversity of fish, with an estimated 2000 to 3000 species present, of which only a minority are exploited commercially (Goulding, 1980; Smith, 1981). Especially for poor people, fish are an important source of nutrition. Amazonia offers great potential for aquaculture, provided fish species are not made extinct through overfishing, pollution or wetland degradation before they can be cultivated. Unfortunately, little is known about the productivity and vulnerability of Amazonian fisheries.

In recent years, river-dwellers along the Brazilian Amazon have tended to shift from diverse, small-scale fishing and other activities to commercial fishing. This has been in response to several factors, including the expansion of ranching on to the *várzea*, the decline of jute cultivation, and improved possibilities for the transport and sale of fish. With better boats and new types of nets have come expanded-plastic cool-boxes and refrigeration systems, which allow fishermen to range further from the packing stations and markets. Manaus has a large commercial fleet that operates up to 1700 km from the city and which often comes into conflict with local fishermen (Serrão, 1994).

Species that are commonly caught include tucunare (*Cichla ocellaris*), pirarucú (*Arapaima gigas*), aruana (*Osteoglossum biccirhosum*), filhote (*Brachyplatystoma filamentosum*), dourado (*B. favicans*), tambaqui (*Colossoma macropomum*) and piranha (*Serrasalmus* spp.). The best species, like pirarucú, have been heavily fished to feed local populations, and they are also frozen for sale beyond Amazonia (Goulding, 1981). Prawns, especially *Macrobrachium amazonicum*, are also an important source of income, particularly in the lower Amazon. Hitherto, they have been netted or trapped in wooden fish traps, but in Ecuador, Peru and elsewhere giant tiger prawns (*Penaeus monodon*) are produced in ponds, whose effluent discharges can have serious impacts on wildlife and fisheries (Primavera, 1991).

A number of Amazonian rivers have been impounded, and there are plans for further barrages. Fisheries have developed in the Tucuruí reservoir on the Tocantins river and now yield more than 3000 t yr^{-1}, much of it highly prized tucunare (*C. ocellaris*) (Boonstra, 1993). Balbina reservoir, near Manaus, also has a productive tucunare fishery. In the Curuá-Una reservoir, near Santarém, the fishery has been less successful as a result of excessive numbers of small piranha (*Serrasalmus* spp.) and parasitic infection of other species (Figure 14.1).

Dam Building

Amazonian rivers have been harnessed to generate power for Manaus and Belém and for the industries and people of the northeast and centre-south of Brazil. Hydroelectric generation is also intended to encourage mineral exploitation in Amazonia by providing cheap power for mining, transport and treatment of ores.

One of the first hydroelectric projects in Amazonia was the massive Tucuruí dam, which, since 1984, has provided power for mining at Serra dos Carajás and for ore smelting and other purposes at Belém. The Tucuruí reservoir flooded some 2100 km^2 of wetland. A further 2430 km^2 were flooded by the Balbina reservoir near Manaus (Fearnside, 1989).

Proposals have also been made to construct small hydroelectric dams (less than 500 kw) on smaller Amazonian streams to serve village communities (Nogueira *et al.*, 1993). Such dams could reduce local fuelwood demand and diminish air pollution from diesel generator exhausts. Small hydroelectric schemes lessen the need for large dams and lengthy cut-lines for power transmission. However, there is still the possibility of damage to some aquatic organisms.

Mineral Exploitation

Much Amazonian mining involves dredging of alluvial sediments, and this affects both floodplains and riverine systems. There is large-scale commercial mining of bauxite, copper, kaolinite and iron ore and small-scale gold mining. There has been a gold rush in Amazonia since the late 1970s, when gold prices rose significantly, and there are currently an estimated 650 000 small-scale gold-diggers (*garimpeiros*) in Brazilian Amazonia (Smith *et al.*, 1991; Greer, 1993). As far as large-scale mining is concerned, the Brazilian government is promoting this as a major component of its long-term development strategy for the region.

Impacts on Riverine Ecosystems

Declining Amazonian fish and prawn stocks cause concern (Chapman, 1989), and it is vital that the reasons for the decline are better understood. Fishing itself is only partly responsible for the decline, which appears also to reflect increasing exploitation of the *terra firme*, modified river flows causing siltation and pollution, increased exploitation of the *várzea*, and hydroelectric developments. The expansion of commercial fishing and the break-down of local traditions that previously discouraged overfishing have been particularly damaging (Smith, 1981). There are now increasing calls for enforced, legal regulation of the industry by means of fishing permits, fish quotas and the like, so that fish stocks can be conserved. Some communities are seeking direct control of local *várzea* lake fisheries in order to improve fish management and sustain long-term production (McGrath *et al.*, 1993).

There is potential for improving Amazonian fisheries and for aquaculture, but any developments should be carefully managed to avoid environmental damage (De Merona, 1990; Flores *et al.*, 1990; McGrath *et al.*, 1993). In particular, it is unwise to treat agriculture, fisheries and other extractive activities as separate issues, given the way they interact (Serrão, 1994).

The impact of dams in Amazonia has been significant. Water released downstream may be anoxic, acidic and/or contaminated with toxic compounds, and, unless it is discharged with caution, can injure fish and other aquatic life (Fearnside, 1989). Siltation upstream from dams affects the food supply of aquatic life downstream (Magee, 1989), and modifies depositional regimes on the *várzea*. Prawn fisheries have been damaged below the Tucuruí dam and the livelihood of many river people has been affected (De Merona *et al.*, 1987; Odinetz-Collart, 1987). The migration of fish is influenced by dams (Reeves and Leatherwood, 1994), and reduced flow variation and flooding impair feeding and jeopardize breeding. Commercial species that are affected include jaraqui (*Semaprochilodus* spp.) and dourado (*Brachyplatystoma favicans*). Loss of fish species was reported following construction of the Tucuruí and Samuel dams (Smith *et al.*, 1991). Similar impacts have been caused by dam-building along the San Francisco river in Bolivia and the Guapore river in Rondônia, Brazil (Diegues, 1992).

In Rondônia, tin mining has caused siltation of streams (Smith, 1981). Elsewhere, large-scale mining has resulted in the discharge of debris and toxic wastes which disperse through the aquatic system and associated wetlands. Aluminium and iron-ore processing in eastern Amazonia causes air pollution resulting in acid deposition over wide areas (Anderson, 1990b). This damages vegetation and releases toxic elements in the soils.

Mercury is used by *garimpeiros* in sluices to trap gold particles. It may be discharged directly into streams or escape as fumes during separation of the gold. The fumes poison miners through inhalation, and they also enter the riverine and wetland food-web. Most of the gold produced in Brazilian Amazonia is sold through unofficial dealers, so it is difficult to make accurate estimates of production, and hence of pollution levels. Malm *et al.* (1990) estimate that over 100 t of mercury entered the Madeira river between 1979 and 1985. Substantial amounts have also been added to other rivers, notably the Branco, Tapajós, Garupi and Tocantins, where *garimpeiros* are active (Cleary, 1990; Smith *et al.*, 1991; Greer, 1993).

Along these channels, samples of hair from people in riverside dwellings and from fish-tissue samples show high levels of mercury, often well above WHO safety limits (Martinelli *et al.*, 1988; Pfeiffer *et al.*, 1991; Nriagu *et al.*, 1992; Cleary *et al.*, 1994; Aks *et al.*, 1995). Mercury accumulates in organisms and threatens the health and long-term survival of fish, wildlife and humans. Greer (1993) describes mercury as 'the pauper's poison' because *garimpeiros* are particularly exposed to the fumes, while many other people living near rivers suffer because they cannot afford to switch from fish to other sources of protein.

The long-term effects are even more serious. Once it is trapped in sediments, toxic mono-methyl mercury will be released from the riverbed and *várzea* deposits by microbial action for decades, even though the mercury use itself is curtailed (Martinelli *et al.*, 1988; Greer, 1993). Techniques for replacing or reducing the amount of mercury used in gold recovery are being investigated, but no simple or economic solutions are immediately available (Cleary, 1990; Coghlan, 1994). Given the large profits being made by the 'gold barons', the apparent laundering of drug profits through gold, and the poverty and lack of alternative opportunities for *garimpeiros*, the control of mercury pollution and its environmental and social impacts in Amazonia is exceedingly difficult.

Other Impacts on Wetland and Riverine Environments

Urban–Industrial Developments

Much of Amazonia's population lives in cities located beside rivers, and their sewage, industrial pollutants and vehicle emissions threaten large tracts of wetland and adjacent lakes and rivers. In particular, fisheries near Manaus and Belém have already been damaged in this way.

There is a variety of other industrial activities that have the potential for serious environmental damage. The cutting of forest to provide charcoal for iron-ore smelting in the Carajás region in eastern Pará is likely to cause soil erosion and acid deposition, both of which will damage wetlands and rivers. Acid deposition is particularly harmful to soils and water bodies which are already naturally acidic. Wood-pulp production is currently being undertaken on the Jari river in northern Pará, where both natural forest and plantation timber supply the raw material (Hoppe, 1992). Pulp production can cause both air and river pollution (Halperin, 1980). Small rum distilleries along the lower Tocantins river are also a source of local pollution.

Agriculture and Ranching on the *Terra Firme*

Forest clearance on the *terra firme* increases surface water run-off with direct effects on silt load and the whole fluvial regime. Such changes damage fisheries and aquatic wildlife. Settlers in Amazonia are increasingly using agro-chemicals which can pass into wetlands, lakes and rivers, causing weed and algal blooms and toxicity in aquatic organisms.

A particular hazard is the commercial production and processing of coca (*Erythroxylum* spp.) in the upper Amazon region, notably in Peru, Bolivia and Colombia. More intensive cultivation in recent years has increased run-off and siltation in rivers, while the processing of coca in forest 'factories' has caused serious river pollution. An estimated 38 000 t yr^{-1} of toxic wastes, including kerosene, sulphuric acid, acetone, toluene and toilet paper (used for filtering), have been entering Amazonian headwaters, with unspecified, but potentially damaging, effects on fish and other aquatic organisms. Attempts to discourage coca cultivation have led anti-drug squads to apply herbicides like 2,4-D and tebuthurion; the latter are likely to remain in aquatic environments for up to five years, killing birds, fish and other animals (Redclift and Sage, 1994).

Changing Sea Level

Many wetlands in eastern Amazonia are subject to tidal inundation or lie at or close to sea level. Even a slight rise in sea level, associated with global warming, would put such land at risk. Periods of inundation, the extent of flooding, the nature of sediment deposition, and the level of salinity in soils and ground water could all change. Many *caboclos* may adapt to the changes, given that they already cope with considerable year-to-year fluvial variations, but larger-scale commercial operations in the zone are likely to be more vulnerable.

Conclusion

It seems inevitable that development activities associated with both wetland and riverine environments in Amazonia will result in physical damage and loss of biodiversity. Adverse social and other effects on human populations, both rural and urban, may result. Land development in the zone should thus only occur after a careful assessment of specific, potential environmental and socio-economic impacts. In addition to the possibility for major agricultural developments, the Amazonian wetlands offer opportunities for (a) other forms of direct usage, notably grazing and fuelwood collection, (b) for non-consumptive use, including tolerant forest management and tourism, and (c) indirect use, including fish feeding and breeding, storm protection, and flood mitigation. The area can also serve as a reservoir of biodiversity, if national and world governments recognize its intrinsic worth and acknowledge the moral obligations to conserve nature (Barbier, 1993).

Many policy implications derive from the above. They include the need (a) to ensure that adequate, carefully selected wetlands are set aside and properly protected, (b) to control wetland agricultural development, especially when it involves significant use of agro-chemicals, (c) to monitor and control wetland logging activities, (d) to improve the monitoring and management of fisheries in the region, (e) to seek ways to control mercury pollution from gold mining, (f) to control impacts associated with the drug trade, and (g) to monitor and control urban–industrial pollution affecting the wetland zone.

Wetland conservation requires the establishment of extensive reserves and riverside belts which are secure from external impacts like dam building and sea-level change. Such areas will provide biological and physical protection for the environment. In some areas of this kind, careful management will allow a degree of exploitation, involving tolerant forest management or controlled logging, as well as conservation. Elsewhere, strict conservation is required. In places, more intensive development is inevitable and permissible. What is needed above all is an integrated approach to the planning and management of Amazonian wetlands and the associated riverine environments.

References

Aks, S.E., Erickson, T.B., Branches, F.J.P. and Hryhorczuk, D.O. (1995) Blood mercury concentrations and renal biomarkers in Amazonian villagers. *Ambio*, **24**, 103–5.

Alcorn, J.B. (1990) Indigenous agroforestry systems in the Latin American tropics. In M.A. Altieri and S.B. Hecht (eds), *Agroecology and Small Farm Development*. Boca Raton: CRC Press, pp. 203–13.

Anderson, A.B. (1990a) Extraction and forest management by rural inhabitants in the Amazon estuary. In A.B. Anderson (ed.), *Alternatives to Deforestation: Steps toward Sustainable Use of the Amazon Rain Forest*. New York: Columbia University Press, pp. 65–85.

Anderson, A.B. (1990b) Smokestacks in the rainforest: industrial development and deforestation in the Amazon Basin. *World Development*, **18**, 1191–205.

Anderson, A.B. (1992) Land-use strategies for successful extractive economies in Amazonia. In D.C. Nepstad and S. Schwartzman (eds), *Non-timber Products from Tropical Forests: Evaluation of a Conservation and Development Strategy. Advances in Economic Botany No. 9*. New York: New York Botanic Garden, pp. 67–77.

Anderson, A.B., Gely, A., Strudwick, J., Sobel, G.L., Das Cracas, C. and Pinto, M. (1985) Uma sistema agroflorestal na várzea do estúario amazônica (Ilha das Onças, Município de Barcarena, Estado do Pará). *Acta Amazônia*, **15**, 195–224.

Anderson, A.B. and Jardim, M.A. (1989) Costs and benefits of floodplain management by rural inhabitants in the Amazon estuary: a case study of acai palm production. In J.O. Browder (ed.), *Fragile Lands of Latin America: Strategies for Sustainable Development*. Boulder, CO: Westview Press, pp. 114–29.

Barbier, E.B. (1993) Sustainable use of wetlands. Valuing tropical wetland benefits: economic methodologies and applications. *Geographical Journal*, **159**, 22–32.

Barrow, C.J. (1990) Environmentally appropriate, sustainable small-farm strategies for Amazonia. In D. Goodman and A. Hall (eds), *The Future of Amazonia: Destruction or Sustainable Development*. London: Macmillan, pp. 153–84.

Boonstra, T.E. (1993) Commercialization of the Tucurui Reservoir fishery in the Brazilian Amazon. *TCD Newsletter, Center for Latin American Studies, University of Florida*, **28**, 1–4.

Chapman, M.D. (1989) The political ecology of fisheries depletion in Amazonia. *Environmental Conservation*, **16**, 331–7.

Cleary, D. (1990) *Anatomy of the Amazon Gold Rush*. London: Macmillan.

Cleary, D. (1991) The greening of the Amazon. In D. Goodman and M. Redclift (eds), *Environment and Development in Latin America: The Policies of Sustainability*. Manchester: Manchester University Press, pp. 116–40.

Cleary, D., Thornton, I., Brown, N., Karantzis, G., Delves, T. and Worthington, S. (1994) Mercury in Brazil. *Nature*, **369**, 613–14.

Coghlan, A. (1994) Midas touch could end Amazon's pollution. *New Scientist*, **141** (1916), 10.

De Merona, B. (1990) Amazonian fisheries – general characteristics based on two case studies. *Interciencia*, **15**, 461–8.

De Merona, B., De Carvalho, J.L. and Bittencourt, M.M. (1987) Les effets immédiats de la fermature du barrage de Tucuruí (Brésil) sur l'ichtyofaune en aval. *Revue de Hydrobiologie Tropicale*, **20**, 73–84.

Denevan, W.M. (1966) A cultural–ecological view of the former aboriginal settlement in the Amazon region. *Professional Geographer*, **18**, 346–51.

Diegues, A.C.S. (1992) Sustainable development and peoples' participation in wetland ecosystem conservation in Brazil: two comparative studies. In D. Ghai and J.M. Vivian (eds), *Grassroots Environmental Action: Peoples' Participation in Sustainable Development*. London: Routledge, pp. 141–58.

Eden, M.J. (1990) *Ecology and Land Management in Amazonia*. London: Belhaven.

Eden, M.J. and Andrade, A. (1988) Colonos, agriculture and adaptation in the Colombian Amazon. *Journal of Biogeography*, **15**, 79–85.

FAO (1993) *Management and Conservation of Closed Forests in Tropical America. FAO Forestry Paper No. 101*. Rome: Food and Agriculture Organization.

Fearnside, P.M. (1989) Brazil's Balbina Dam: environment versus the legacy of the pharaohs in Amazonia. *Environmental Management*, **13**, 401–23.

Flores, H.G., Bocanegra, F.A., Garcia, J.M. and Riveiro, H.S. (1990) Fisheries in the Peruvian Amazon. *Interciencia*, **15**, 469–75.

Goodland, R.J. and Irwin, H.S. (1975) *Amazon Jungle: Green Hell to Red Desert?* Amsterdam: Elsevier.

Goodman, D. and Hall, A. (eds) (1990) *The Future of Amazonia: Destruction or Sustainable Development*. London: Macmillan.

Goodman, D. and Redclift, M. (eds) (1991) *Environment and Development in Latin America: The Policies of Sustainability*. Manchester: Manchester University Press.

Goulding, M. (1980) *The Fishes and the Forest. Explorations in Amazonian Natural History*. Berkeley: University of California Press.

Goulding, M. (1981) *Man and Fisheries on an Amazon Frontier*. The Hague: W. Junk.

Gray, A. (1990) Indigenous people and the marketing of the rainforest. *Ecologist*, **20**, 223–7.

Greer, J. (1993) The price of gold: environmental costs of the new gold rush. *Ecologist*, **23**, 91–6.

Grenard, F. and Grenard, P. (1993) Historical stages of the várzea settlement in the Amazon. *Amazoniana-Limnologia et Oecologia Regionalis Systemae Fluminis Amazonas*, **12**, 509–26.

Hall, A. (1989) *Developing Amazonia: Deforestation and Conflict in Brazil's Carajas Programme*. Manchester: Manchester University Press.

Halperin, D.T. (1980) The Jari Project: large-scale land and labor utilization in the Amazon. *Geographical Survey*, **9**, 13–21.

Hiraoka, M. (1993) Sustainable resource management in the Amazon floodplain: report on the first stage of the 'Varzea Project'. *PLEC News and Views*, Canberra, **1**, 11–13.

Hoppe, A. (1992) The Amazon between economy and ecology. *Natural Resources Forum*, **16**, 232–4.

Kahn, F. and de Granville, J-J. (1992) *Palms in Forest Ecosystems of Amazonia*. Berlin: Springer-Verlag.

Lima, R.R. (1956) A agricultura nas várzeas do estuário do Amazonas. *Boletim Técnico, Instituto Agronômico do Norte*, Belém, **33**, 1–164.

Lopez-Parodi, J. and Freitas, D. (1990) Geographical aspects of forested wetlands in the lower Ucayali, Peruvian Amazonia. *Forest Ecology and Management*, **33–4**, 157–68.

McGrath, D.G., De Castro, F., Futemma, C., de Amaral, B.D. and Calabria, J. (1993) Fisheries and the evolution of resource management on the lower Amazon floodplain. *Human Ecology*, **21**, 167–96.

Macedo, D.S. and Anderson, A.B. (1993) Early ecological changes associated with logging in the lower Amazon floodplain. *Biotropica*, **25**, 151–63.

Madeley, J. (1993) Raising rice in the savannas. *New Scientist*, **138** (1878), 36–9.

Magee, P. (1989) Peasant political identity and the Tucuruí Dam: a case study of the island dwellers of Pará, Brazil. *Latinamericanist*, **24**, 6–10.

Malm, O., Pfeiffer, W.C., de Souza, C.M.M. and Reuther, R. (1990) Mercury pollution due to gold mining in the Madeira River basin, Brazil. *Ambio*, **19**, 11–15.

Martinelli, L.A., Ferreira, J.R., Fosberg, B.R. and Victoria, R.L. (1988) Mercury contamination in the Amazon: a gold rush consequence. *Ambio*, **17**, 252–4.

Nogueira, M.F.M., Lima, C.U.D.S. and Ribeiro, R.R.P. (1993) The use of small hydroelectric power plants in the Amazon. *Renewable Energy*, **3**, 907–11.

Nriagu, J.O., Pfeiffer, W.C., Malm, O. and de Souza, C.M.M. (1992) Mercury pollution in Brazil. *Nature*, **356**, 389.

Nugent, S. (1993) *Amazonian Caboclo Society: An Essay on Invisibility and Peasant Economy*. Berlin: Berg.

Odinetz-Collart, O. (1987) La pêche crevettère de *Macrobrachium amazonicum* (Palaemonidae) dans le Bas-Tocantins, après la fermature du barrage de Tucuruí (Brésil). *Revue de Hydrobiologie Tropicale*, **20**, 131–44.

Pfeiffer, W.C., Malm, O., de Souza, C.M.M., Drude de Lacerda, L., Silveira, E.G. and Bastos, W.R. (1991) Mercury in the Madeira River ecosystem, Rondonia, Brazil. *Forest Ecology and Management*, **38**, 239–45.

Prance, G. (1990) Future of the Amazonian rainforest. *Futures*, **22**, 891–903.

Primavera, J.H. (1991) Intensive prawn farming in the Philippines – ecological, social and economic implications. *Ambio*, **20**, 28–33.

Redclift, M. and Sage, C. (eds) (1994) *Strategies for Sustainable Development: Local Agendas for the Southern Hemisphere*. Chichester: John Wiley and Sons, pp. 171–86.

Reeves, R.R. and Leatherwood, S. (1994) Dams and river dolphins: can they co-exist? *Ambio*, **23**, 172–5.

Roosevelt, A. (1989) Lost civilisations of the lower Amazon. *Natural History*, **98**, 74–83.

Serrão, E.A. (1994) The Amazon floodplain: the next major frontier for food production. *PLEC News and Views*, Canberra, **2**, 25–8.

Smith, N.J.H. (1981) *Man, Fishes and the Amazon*. New York: Columbia University Press.

Smith, N.J.H., Alvim, P., Homma, A., Falesi, I. and Serrão, A.E. (1991) Environmental impacts of resource exploitation in Amazonia. *Global Environmental Change*, **1**, 313–20.

15

Degradation of Mangrove Forests Adjacent to Urban Areas in Belize and Fiji: A Comparative Study

Frank McShane

Mangroves are complex plant communities that occur within the inter-tidal zone in the tropics. Their function and utility are well-documented (e.g. Odum and Heald, 1972; Tomlinson, 1986; Bastian and Benfarado, 1988; Altenberg and Van Spanje, 1989; Steinke and Ward, 1989), but, in the vicinity of urban areas, they are frequently under pressure and even threat of conversion to alternative uses (Jayapaul *et al.*, 1988; Hogarth, 1989; Fortes, 1991). Coastal peoples living in such areas and having limited income-earning opportunities commonly harvest mangrove products to supplement their diet or income, while the wider urban community benefits from various mangrove functions. For such groups, mangrove degradation has serious impacts. In this chapter, data from Belize City in Central America and Suva in Fiji exemplify the function and utility of mangroves in the vicinity of urban areas and the development pressures to which they are subject.

Development pressures on mangroves in Belize as a whole are not severe, but mangrove degradation is increasingly taking place in urban areas, particularly Belize City. The population of the city has increased by some 16 per cent in the last decade to reach 46 020 (Department of Statistics, 1991). Poor planning in the past has led to urban overcrowding and associated social and health problems that are exacerbated by inadequate sewage disposal. Rapid urban growth and poor planning have increased the pressure on mangroves.

In Fiji, the mangroves areas around Suva are smaller, more dispersed and less degraded than those around Belize City. Although the population of Suva is larger (about 120 000 in greater Suva), indigenous control of mangrove-related fish resources has limited the cutting of mangrove. In Fiji, compensation is also paid for loss of access to traditional fish resources as a result of mangrove clearance. While similar degradation is evident in both places, management strategies vary as a result of the societal contrasts.

Mangroves of Belize with Emphasis on the Belize City Area

Belize City lies on the east coast of Central America. The area has a humid subtropical climate, with mean annual rainfall in the range 510–1370 mm (Hartshorn *et al.*, 1989). The coast itself is at or below sea level, which has created extensive wetlands that include Fabers and Jones lagoons. These are linked to the Northern and Southern lagoons further south of the city. These areas are tidal, and their dominant vegetation is mangrove forest (Figure 15.1).

The mangroves of Belize City are characterized by four species, which are distinguished by their morphology, salinity tolerance and habitat. *Rhizophora mangle* is dominant and encountered in frequently inundated areas. *Avicennia germinans* tolerates higher salinity than *Rhizophora*, while *Laguncularia racemosa* prefers lower salinity and higher ground. *Conocarpus erectus* has the lowest salinity tolerance and was previously thought to be a mangrove-associate rather than a true mangrove (Cintron and Schaeffer-Novelli, 1983). The term mangrove is also applied to the complex plant communities associated with the above species. The area around Belize City contains examples of the main mangrove communities identified in the country, namely, fringing mangrove forest, saltmarsh with sparse mangrove, basin mangrove forest and riverain mangrove forest (Zisman, 1990).

The Importance of Mangroves in the Belize City Area

Mangroves provide significant direct-use products and have an important function and utility for the urban and peri-urban population of the city.

1 Direct-use products
Around Belize City, mangrove wood is used for cooking, burning and making charcoal, particularly in squatter settlements. The wood is also used for flooring, panelling and fence posts. Many other uses are reported in Belize by Furley and Ratter (1992), but were not encountered in the urban area.

2 Habitat functions
Various commercial fish species are associated with mangrove in Belize (Barrick, 1989). Tarpon snook (*Centropomus pectinatus*), various species of the Serranidae, and several species referred to as 'snapper' are harvested for

subsistence use. Penaeid shrimp species are fished for subsistence and commercially, particularly at Jones Lagoon and the mouth of the Sibun river. White shrimp (*Penaeus schmitti*) is a near-shore species (Longhurst and Pauly, 1987) and an important subsistence resource. In Malaysia (Chong *et al.*, 1990), the Philippines (Martsubroto and Naamin, 1977) and Australia (Staples,

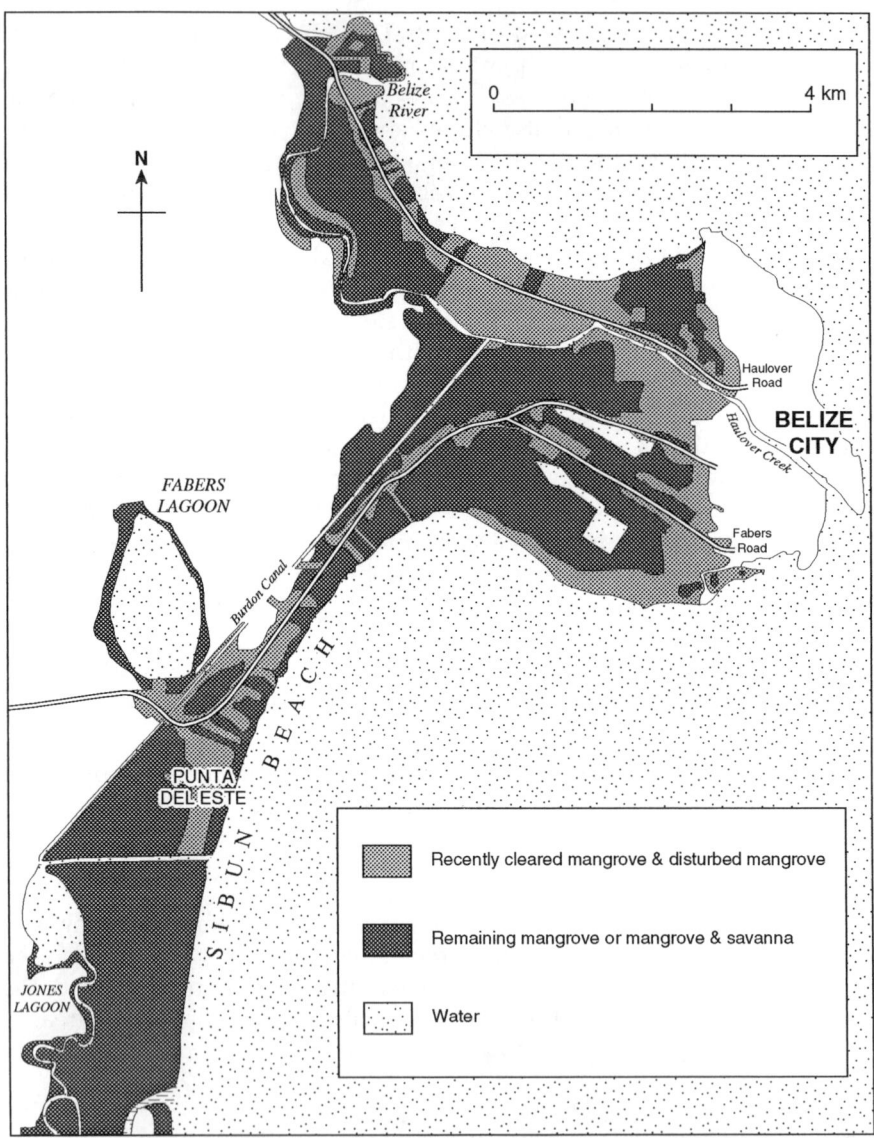

Figure 15.1 Distribution of mangroves around Belize City, Belize.

1985), a clear relationship has been demonstrated between shrimp abundance and mangrove area.

Mangroves provide a sheltered habitat for many crab species, and some decapod crustacea utilize mangrove leaves for transport in the juvenile stages, reducing predation and energy loss (Wehrtmann and Dittel, 1990). Crabs in turn provide a food source for predators like hawks and herons and aid the redistribution of sediment through the turnover of soil (Walcott, 1988). Around Belize City, the fiddler crab (*Uca* sp.) and blue land crab (*Cardisoma guanhumi*) are particularly abundant in the *Rhizophora mangle* mangrove around the Belize river mouth and on shores to the south. Small specimens of the mangrove crab (*Aratus pisonii*) were also encountered on the upper roots of *Rhizophora mangle*. Mangrove roots also host a rich assemblage of sessile organisms, particularly molluscs, algae and bacteria that may be a significant nutrient source in mangrove areas (Morton, 1983; Rodriguez and Stoner, 1990). The mangrove oyster (*Crassostrea rhizoporae*) is also harvested in the area, notably on the southern shore (McShane, 1991).

Mangrove communities attract insectivorous, piscivorous and crab-eating waders and water fowl (Cawkell, 1964; Altenberg and Van Spanje, 1989). Few bird studies have been done in Belizean mangroves; the present author observed cattle egrets, ibis, tanager, heron, kingfisher and northern jacana in and around the lagoons south of Belize City. Further south on Western Highway, sightings were made of humming-birds and roadside hawk. Along the Belize river, toucan, aracari as well as tanagers and fly-catchers were sighted in riverain mangrove and tall mangrove at Tillets pond. Pelicans and frigate birds are frequently seen diving for fish close to mangrove-fringed shores.

Belize mangroves provide a habitat for the West Indian manatee (*Trichechus manatus*), which is listed as an endangered species by the International Union for the Conservation of Nature and Natural Resources. The lower Belize river, including the city area, recorded the second highest count of manatees in a national survey conducted in 1989 and the population status in the country is described as unique (O'Shea and Salisbury, 1991). Morlet's crocodile (*Crocodylus moreleti*) and American crocodile (*C. acutus*) also occur in the coastal, lagoon and river waters of the Belize City area.

3 Coastal protection functions

Tropical cyclones are often accompanied by tidal surges that damage low-lying coasts (Linden and Jernalov, 1980). Between 1931 and 1985, eighteen hurricanes affected Belize, the most severe being Hurricane Hatty on 31 October 1961. The storm was followed by a 3 m tidal surge that caused extensive damage around Belize City (British Honduras Hurricane Assessment Mission, 1962). Mangroves reduce such storm effects by limiting wave amplitude and velocity as waves travel through them. The closer the trees are together, the greater the attenuation (Othman, 1994).

4 Water quality control functions
According to Bastian and Benfarado (1988), wetlands improve the quality of terrestrial run-off through filtering. The retention of waste water in wetlands achieves more efficient aerobic breakdown of bacterial matter, which aids the control of human pathogens in waste water discharge. In the case of Belize City, sewage reservoirs to the south are located within the coastal lagoon system. Sewage disposal from the primary settling pits is through an over-flow to the coast. The high density of mangrove pneumatophores in the lagoon area increases waste water retention in the system and improves coastal water quality.

5 Recreation and tourism
The mangrove-fringed Burdon canal and Haulover creek offer unique recre-ational opportunities. The Burdon canal connects Belize City with the ex-tensive lagoon system to the south. Built in the 1920s for navigation, the canal is one of the few in the area not chronically polluted by sewage dumping. The canal is nowadays only used by small boats, but offers easy access to Bird Cay Nature Reserve in Northern lagoon. The reserve, a rookery for egrets, boat-billed herons, anhingas and other species, attracts tourists. If properly man-aged, tourism could provide an alternative income for local people.

Degradation of Mangroves around Belize City

Recent legislation requires a permit to cut mangrove (Government of Belize, 1989), but illegal cutting continues. In 1939, some 90 per cent of the mangrove forest around Belize City remained in place, but, by 1991, only 52 per cent survived. Between 1988 and 1991, the rate of clearance averaged 3.6 per cent per annum (McShane, 1991). The main causes of clearance are land reclama-tion for housing, industrial and commercial purposes, garbage disposal and beach recreation. Clearance along shores and river banks has increased erosion, necessitating engineering works particularly along the Northern Highway.

Along Haulover creek, mangrove clearance has caused significant soil erosion and raised sediment load in the waterway. Clearance by 'home-steaders' near the sewage lagoons towards Belizean Beach has also caused erosion. In addition, trimming of mangrove roots for fuelwood by home-steaders reduces the retention time of waste water in lagoons and causes tree dieback. This affects coastal water quality and is a potential health risk when the waste water overflows into the lagoon.

Further damage is likely to occur to the mangrove at the mouths of the Belize river, Sibun river and Boom creek where land has been designated for private sector development. The Burdon canal area is also under threat from piecemeal commercial developments along the Western Highway, while the nature reserve at Fabers lagoon has been partly cleared for hotel development. Extension of the city boundaries has increased the pressure on mangrove forests. Two private housing projects, Punta del Este (250 ha) at Sibun Beach and Vista del Mar (120 ha) at Ladyville, have involved extensive mangrove clearance. No prior environmental impact assessments were made, while both projects offer housing that is too expensive for lower-income Belizeans who are most in need of rehousing (McShane, 1991).

Mangrove Management in Belize

The main legislation for managing mangroves in Belize is the Forests (Protection of Mangroves) Regulations (1989). The Department of Forestry issues permits and levies fines for mangrove cutting. In addition, the Ministry of Tourism and Environment co-operates with other public agencies that have interests in mangrove management, notably the Fisheries Department and the Ministry of Economic Development.

Other legislation with implications for mangrove management includes the National Parks Systems Act (1981) and the Natural Resources and Wildlife Protection Act (1981). Land laws, including the Crown Lands Ordinance, the Land Tax Act, and the Land Utilization Act, also affect mangrove conservation. The Land Tax Act, for example, allows taxes to be levied on privately owned land, based on its unimproved value, which in effect provides an incentive to clear mangroves.

The Belize City Council extended the city boundaries in 1991 on the basis of a zoning and development plan from the Belize City Housing and Planning Unit (1991). The plan recommended the conservation of productive mangroves within the city limits and the maintenance of coastal and riverain mangrove buffer zones. Protected areas include the sewage lagoons, Haulover creek, Burdon canal and the coastal strip.

Mangroves of Fiji with emphasis on Suva Peninsula and Rewa delta

Fiji is an archipelagic state of over 300 islands in the southwest Pacific. Much of its area of 18 300 km² consists of the islands of Viti Levu (10 544 km²) and

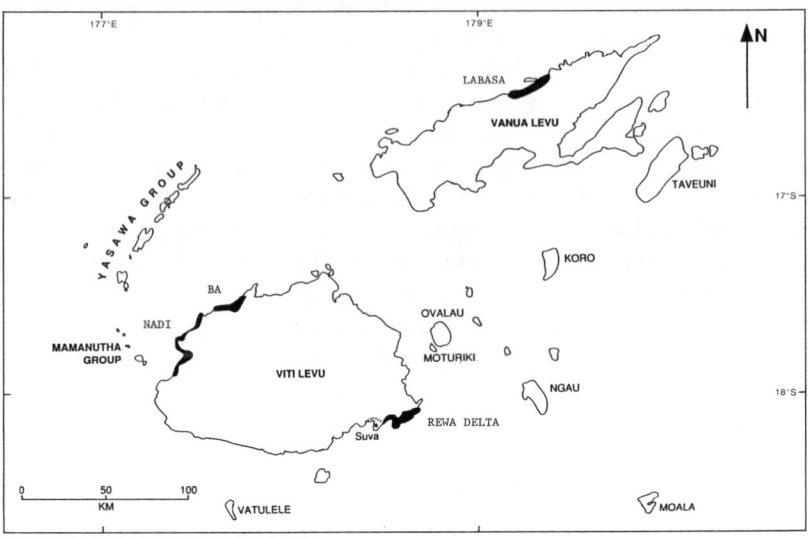

Figure 15.2 Areas of extensive mangrove cover in Fiji. *Source:* after Gray (1993).

Vanua Levu (5535 km²). Suva, the capital of Fiji, is located on Viti Levu (Figure 15.2). The islands have a tropical maritime climate, with rainfall concentrated in the period November to April. The eastern side of Viti Levu, which is exposed to the southeast trades, has an annual rainfall of 1800 to 2600 mm, with 1300 to 1600 mm on the western side.

The coasts of the high volcanic islands provide the most extensive mangrove habitat. Watling and Chape (1992) estimate the current area of mangrove at 42 000 ha, as against an original cover of 45 000 ha. The largest concentrations are found at the mouths of the Ba, Rewa, Nadi and Labasa rivers.

The mangrove communities of Fiji appear to have a relatively simple floristic composition, but this may reflect inadequate collection and taxonomic analysis. Gray (1993) records nine mangrove species, with *Rhizophora* spp. (*R. samoensis, R. stylosa, R. x selala*) dominant in most communities; other species include *Bruguiera gymnorrhiza, Exocoecaria agallocha* and *Xylocarpus granatum*. Vodonaivalu (1983) identifies 86 related mangrove species consisting of 'shrubs or trees and many epiphytes both of ferns and orchids'.

The largest and most diverse mangrove community in Fiji lies east of the Suva peninsula in the Rewa delta (Figure 15.2). The delta contains over 100 village communities that supply Suva with local products. The mangrove communities include extensive areas of *Bruguiera gymnorrhiza* closed forest and of mixed open forest with *B. gymnorrhiza* and *Rhizophora x selala*; subsidiary areas exist of shrub forest, mixed fringing forest and mixed closed forest (Watling, 1985).

The Importance of Mangroves in the Rewa Area

Mangroves in Fiji provide a means of supplementing formal incomes and provide resources of both subsistence and commercial value.

1 Direct use products

The Rewa delta mangroves are a valuable source of domestic fuelwood, yielding an estimated 5000 t yr^{-1} (Watling, 1985). This level of exploitation is thought to be sustainable, but the local population is increasing and illegal wood cutting occurs. Mangrove charcoal is also occasionally produced for urban barbecues (Lal, 1991). While commercial cutting of mangroves is controlled by licensing, subsistence cutting is not controlled and estimates of the rate and use of the product are lacking. However, cutting does not generally occur without the consent of the owners of traditional fishing rights.

2 Habitat functions

For the Rewa delta mangroves, Raj *et al.* (1984) list seven fern species, twenty-six families of flowering plants, and numerous molluscs, crustacea, echino-derms, insects, annelids, reptiles and birds. Seventeen fish species have been recorded, including rabbit fish, grunts, goatfish, pony fish and mullet; the recorded total is an underestimate.

   People frequently fish in the mangroves around the Rewa delta. No precise data exist, but Lal (1989) estimated that the total harvest of fish and related products from Fijian mangroves was 7181 t yr^{-1}. According to Gray (1993), more than 60 per cent of Fiji's food fishes utilize the mangroves at some stage of their life history. Much of the near-shore subsistence catch may thus be mangrove-dependent, while some species like mangrove crab are intimately associated with the habitat. According to Raj *et al.* (1984), many species are important sources of protein for people who live near mangroves and river mouths.

3 Coastal protection functions

Fiji is in a cyclone belt and commonly suffers severe flooding and storm surges after cyclones. A 2- to 3-m surge was recorded on offshore islands near Fiji after Cyclone Meli in 1977 and a 3- to 4-m surge after Cyclone Oscar in 1983. Damage from tsunami also occurs. The narrow coastal zone around Suva contains a large human population, and coastal protection would be locally enhanced by a mangrove buffer zone. The degree of protection, while never complete, relates to the width of the zone (Watling, 1986). Rabanal (1981) recommends a buffer zone of 100 to 200 m along the coast.

4 Water quality control functions
Mangroves are utilized to advantage at Raiwaqa, north of Suva, where a large sewerage works discharges into Laucala bay mangroves, via a secondary trickling filter system. The sewerage works at Kinoya employs a similar system, occupying a 6-ha site reclaimed from mangrove.

Pressures on Mangroves in the Suva Area

The mangroves of the Suva peninsula are now severely degraded and only isolated remnants exist. The Queens highway, which runs along the coast of Viti Levu west of Suva, gives access to important mangrove areas and increases the development pressure. Land reclamation is the main threat to the remaining mangroves of the Rewa delta and Suva peninsula. Singh (1990, p. 396) indicates that while 'there is a remarkable acknowledgement among the rural community of the subsistence and commercial products of the mangrove, this appreciation is very little reflected in urban dwellers and developers'. Around the Suva peninsula, mangroves are confined to isolated stands of stunted Rhizophoraceae, particularly *B. gymnorrhiza*, except along major waterways where the cover is more extensive.

Various sources of industrial pollution affect the remaining mangrove stands around Suva harbour, including the cement works discharge at Lami and boat discharges in the docks. The Suva city rubbish dump occupies reclaimed mangrove land on the western margin of the city. No plans have yet been approved for its relocation. Land reclamation along the waterfront continues with developments such as the Rokobili container terminal west of Suva, which is an area of traditional fishing. Dredging at the mouth of the Rewa river for flood mitigation also affects mangroves through siltation when spoil is dumped and watercourses altered.

In the delta zone east of the city, clearance for agriculture poses a major threat to mangroves and one that will grow as the population increases (Gray, 1993). The planned Waidamu and Rewa delta seawalls will increase access to and illegal felling of mangroves (Watling, 1986). An acute shortage of freehold land is already leading to mangrove clearance for squatter settlements (Lal, 1991).

Mangrove Management in Fiji

Until 1975 mangroves were designated as Reserved Forests under the control of the Forest Department (Gray, 1993). Thereafter, mangroves within the city boundaries were administered by city councils.

Traditional fishing rights are recognized in mangrove areas and the owner-ship is codified. The loss of fish stocks or fishing areas through development may attract compensation after independent arbitration. Compensation value is assessed on present and future use and on the productivity of the mangrove resource. Arbitration involves the Fisheries Department, the own-ers of fishing rights, the Native Fisheries Commission and the developer. The Fisheries Ordinance and the Crown Lands Ordinance control activities in respect of commercial fisheries and private development respectively.

A Mangrove Management Committee was established in 1983 to prepare a Mangrove Management Plan for Fiji. The plan involved a zonation scheme for the Rewa delta and Suva–Navua area (Watling, 1985, 1986). The zones include: national reserve areas for amenity, scientific and educational use; resource reserves requiring active conservation; traditional use zones for subsistence; wood production zones; shoreline protection zones; and devel-opment zones for agriculture. Where practical, it was recommended that development or wood production should be targeted at remaining freehold land areas. Development can only occur after the issue of a development lease by the Lands Department in consultation with the Mangrove Manage-ment Committee and other interest groups like the Native Fisheries Commis-sion. When the Mangrove Management Plan was formulated, applications to clear mangrove were increasing, as it was generally regarded as suitable land for development. The absence of a clear policy regarding mangrove clearance has resulted in piecemeal development that threatens the resource (Watling, 1986).

Implications of Mangrove Degradation

In both Belize and Fiji, mangroves are important to the urban and peri-urban communities, particularly with regard to fisheries. Mangrove degradation is mainly due to land reclamation and illegal wood cutting, but differences exist in the relative importance of the factors that promote degradation.

In Belize, the legislation is sufficient to conserve mangrove forests, but inadequate resources and insufficient political will exist to enforce it. In practice, poor urban planning and an absence of ecological planning prevent effective conservation. Poor housing, an increasing population, and a per-ceived shortage of well-drained land encourage mangrove clearance. The Ministry of Environment has also failed to inform developers of their envi-ronmental obligations. A desire by local inhabitants to live on the coastal strip and a lack of public awareness of the value of mangroves has ex-acerbated the problem. Homesteaders often believe that uncleared Crown Land is common property ripe for development. In the past, the lack of compulsory environmental impact assessment and of planning controls on urban drift, squatting and associated unlicensed clearance also inhibited conservation.

Proposals have been made to alleviate urban problems by attracting development to inland growth poles (Fairweather *et al.*, 1989; McGill, 1991), but this does not appeal to most urban dwellers and no incentives are offered to move from the coast. As a result, mangroves continue to be degraded. Only a moratorium on mangrove cutting will conserve them.

In Fiji, the indigenous population relies on the mangroves for many products. The mangrove forest has cultural connotations and is associated with the exercise of traditional skills and the extraction of products used in traditional ceremonies. However, although the government promotes the concept of indigenous landowner rights, the anomalous state ownership of land below the mean high-water mark discourages productive management of mangroves.

Traditional or Alternative Uses in Urban Areas?

Lal (1990, p.2) enquires 'under what conditions should ... mangrove be maintained and managed for *in situ* uses ...? When should it be reclaimed to "create" land for alternative purposes?' In economic terms, if the benefits less the costs of conversion to an alternative use give a greater gain than traditional and commercial use and social benefits in perpetuity, then the land use should change. However, such change should only proceed after proper discussion with the owners and users of the resource.

In neither Belize nor Fiji do the principal users own the resource. In Fiji, where procedures exist to compensate traditional users, Lal (1991, p.4) indicates that 'arbitration awards (for loss of usufruct rights) do not compare even with the minimum estimates of expected loss of social benefits'. In an oft-quoted study, Lal (1989) showed that the net benefit of clearing two mangrove areas in Fiji, one for an irrigated rice scheme (242 ha) and the other for sugar cane (350 ha), was less than if the mangrove were maintained. When the entire stream of benefits from the mangrove was accounted for, a negative net present value for conversion existed, which may apply in many mangrove areas.

Dixon (1989) makes the point that a market price cannot be assigned to some products and services derived from the mangrove, noting that some of these may accrue from areas beyond the mangrove. This is important when considering the economics of mangrove clearance around cities where the local people are still highly dependent on mangroves. Yet it is here that the pressure for mangrove clearance is most intense and that market economics invariably favour clearance. Thus, environmental impact assessments must include existing resource use when evaluating mangroves around cities.

Each case needs to be assessed on its merits, but such assessments, though rarely made, will generally favour conserving mangroves. Compensation regimes also need careful evaluation, since compensation may simply encourage mangrove clearance by substituting cash payments, based on present direct-use value, for the value of future benefits. As in Fiji, even relatively large compensation payments may not account for future benefits lost through mangrove clearance. Compensation may also be unevenly distributed through the community.

Economists and ecologists are increasingly working together within planning frameworks to establish new ways of adding value to existing wetlands, including socio-cultural as well as economic and bio-physical value (Burbridge, 1994). Environmental groups in Belize and Fiji are stressing the inherent value of mangroves. Even so, the value of their alternative use often appears greater, and only direct government intervention seems likely to conserve them. Unless strong conservation policies and effective monitoring and enforcement are established, mangroves around cities will surely be cleared and socio-economic damage inflicted on local peoples dependent on them.

References

Altenberg, W. and Van Spanje, T. (1989) Utilisation of mangroves by birds in Guinea Bissau. *Ardea*, **77**, 57–73.

Barrick, P. (1989) A stationary visual census technique. *Proceedings of an International Coastal Resource Management Workshop, 23–25th August, San Pedro, Ambergris Cay.*

Bastian, R.K. and Benfarado, J. (1988) Water quality functions of wetlands; natural and damaged ecosystems. In D.D. Hook, W.H. McKee, H.K. Smith Jr, J. Gregory, V.G. Burrell, M.R. Devoe Jr, R.E. Sojka, S. Gilbert, R. Banks, L.H. Stolzy, C. Brooks, T.D. Matthews and T.H. Shoar (eds), *The Ecology and Management of Wetlands.* London: Croom-Helm, Vol. 1, pp. 87–97.

Belize City Housing and Planning Unit (1991) *Urban Development Plan for Belize City.* Belize City: Government of Belize.

British Honduras Hurricane Assessment Mission (1962) *British Honduras Assessment Mission: Hurricane Hatty. Report to H.M. Government.* Belmopan: Her Majesty's Publication Office.

Burbridge, P.R. (1994) Integrated planning and management of freshwater habitats including wetlands. *Hydrobiologia*, **285**, 311–22.

Cawkell, E.M. (1964) The utilisation of mangroves by African birds. *Ibis*, **106**, 251–3.

Chong, V.C., Sasekumar, A., Leh, M.U.C. and D'Cruz, R. (1990) The fish and prawn communities of a Malaysian coastal mangrove system with comparisons to adjacent mudflats and inshore waters. *Estuarine and Coastal Shelf Science*, **31**, 703–22.

Cintron, G. and Schaeffer-Novelli, Y. (1983) Mangrove forests: ecology and response to man induced stressors. In J.C. Ogden and E.H. Gladfelter (eds), *Coral Reefs,*

Seagrass Beds and Mangroves: Their Interaction in the Coastal Zone of the Caribbean. Report of a Workshop Held at St Croix, U.S. Virgin Islands, May 1982. Paris: UNESCO Reports in Marine Science, 87–113.

Department of Statistics (1991) *Population Census.* Belmopan: Government of Belize.

Dixon, J.A. (1989) Valuation of mangroves. *Tropical Coastal Area Management,* Manila, **4** (3), 1–6.

Fairweather, P., Brown, A. and Wolfe, D. (1989) *Belize City Comprehensive Development Plan.* Montreal: Urban Planning Department, McGill University.

Fortes, M.D. (1991) Seagrass and mangrove ecosystem management: a key to marine conservation in the ASEAN region. *Marine Pollution Bulletin,* **23,** 113–16.

Furley, P. and Ratter, J. (1992) *Mangrove Distribution, Vulnerability and Management in Central America.* ODA-OFI Forestry Research Programme, Contract No. R.4736, Edinburgh.

Government of Belize (1989) *National Development Plan.* Belmopan: Government Publisher.

Gray, A.J. (1993) Fiji: introduction. In D.A. Scott (ed.), *A Directory of Wetlands in Oceania.* Kuala Lumpur: International Waterfowl and Wetlands Bureau and Asian Wetland Bureau, p. 73.

Hartshorn, G., Nicolait, L. and Hartshorn, L. (1989) *Belize Country Environmental Profile.* Belize City: Richard Nicolait and Associates.

Hogarth, P.J. (1989) Mangroves and development around Xiamen, China. In *Coastal Zone, 89. Proceedings of 6th Symposium on Coastal Zone Management, 11–14th July, Charleston, South Carolina,* Vol. 5, p. 43.

Jayapaul, A., Banth, P., Hilda, A. and Selvam, B. (1988) Impacts of urbanisation on the status of mangrove swamps of Madras. In D.D. Hook, W.H. McKee, H.K. Smith Jr, J. Gregory, V.G. Burrell, M.R. Devoe Jr, R.E. Sojka, S. Gilbert, R. Banks, L.H. Stolzy, C. Brooks, T.D. Matthews and T.H. Shoar (eds), *The Ecology and Management of Wetlands.* London: Croom-Helm, Vol. 2, pp. 220–5.

Lal, P.N. (1989) *Conservation or Reclamation of Mangroves: Economic and Ecological Interactions within Mangrove Ecosystems in Fiji.* Unpublished PhD thesis, Department of Agriculture and Natural Resources, University of Hawaii.

Lal, P.N. (1990) *Conversion or Conservation of Mangroves in Fiji: An Ecological Economic Analysis.* Occasional Paper No. 11, Environment and Policy Institute, East–West Centre, Honolulu.

Lal, P.N. (1991) *Mangrove Management Issues: Strategies Adopted in the Pacific Islands.* Island/Australia Working Paper No. 91/3, Canberra: National Centre for Development Studies, Australian National University.

Linden, O. and Jernalov, A. (1980) The mangrove swamp: an ecosystem in danger. *Ambio,* **9** (2), 81–90.

Longhurst, A.R. and Pauly, D. (1987) *The Ecology of Tropical Oceans.* London: Academic Press.

McGill, J. (1991) *Rural Development Plan for Belize District.* Belize: Rural Planning Department, Ministry of Natural Resources.

McShane, F. (1991) *Environmental and Social Implications of Mangrove Clearance around Belize City.* Unpublished MSc thesis, Heriot-Watt University, Edinburgh.

Martsubroto, P. and Naamin, N. (1977) Relationships between tidal forests (mangroves) and commercial shrimp in Indonesia. *Marine Research in Indonesia,* **18,** 81–6.

Morton, B. (1983) Mangrove bivalves. In W.D. Russel-Hunter (ed.), *The Mollusca. Vol. 6. Ecology*. London: Academic Press, pp. 77–138.

Odum, W.E. and Heald, E. (1972) Trophic analysis of an estuarine mangrove community. *Bulletin of Marine Science*, **22**, 671–738.

O'Shea, T.J. and Salisbury, A.L. (1991) Belize – a last stronghold for manatees in the Caribbean. *Oryx*, **25**, 156–64.

Othman, M.A. (1994) Value of mangroves in coastal protection. *Hydrobiologia*, **285**, 277–82.

Rabanal, H.R. (1981) *The Development of Aquaculture in Fiji*. Bangkok: World Food and Agricultural Organisation.

Raj, U., Vodonaivalu, S., Seeto, J., Hirata, H. and Iwakiri, S. (1984) *Flora and Fauna of Mangroves in the Rewa Delta, Fiji Islands*. Suva: Institute of Marine Resources, University of the South Pacific.

Rodriguez, C. and Stoner, A.W. (1990) The epiphyte community of mangrove roots in a tropical estuary. *Aquatic Botany*, **36**, 117–26.

Singh, B. (1990) The role of mangroves in national development: a case study, Fiji. *Regional Workshop on Environmental Management and Sustainable Development, 17th– 21st April, Suva*, 394–403.

Staples, D.J. (1985) Modelling the recruitment process of the banana prawn, *Penaeus merguiensis* in the south east gulf of Carpentaria. In P.C. Rothliesberg, B.J. Hill and D.J. Staples (eds), *The Second Australian National Prawn Seminar*. Australia: Cleveland, pp. 175–84.

Steinke, T.D. and Ward, C.J. (1989) Some effects of the cyclones Domoina and Imboa on mangrove communities in the St. Lucia estuary. *South African Journal of Botany*, **55**, 340–8.

Tomlinson, P.B. (1986) *The Biology of Mangroves*. Cambridge: Cambridge University Press.

Vodonaivalu, S. (1983) *A Botanical Survey of the Tidal Forests (Mangal) of Fiji, Tonga and Western Samoa*. Suva: Institute of Marine Resources, University of the South Pacific.

Walcott, T.G. (1988) The biology of the land crabs. In W.W. Bruggeren and B.R. McMahon (eds), *The Biology of the Land Crabs*. Cambridge: Cambridge University Press, pp. 55–96.

Watling, D. (1985) *A Mangrove Management Plan for Fiji: Phase 1*. Suva: Government Press.

Watling, D. (1986) *A Mangrove Management Plan for Fiji: Phase 2*. Suva: Government Press.

Watling, D. and Chape, S. (eds) (1992) *Environment Fiji: The National State of the Environment Report*. Gland: International Union for Conservation of Nature.

Wehrtmann, I.S. and Dittel, A.I. (1990) Utilisation of mangrove leaves as a transport mechanism with an emphasis on decapod crustacea. *Marine Ecology Progress Series*, **60**, 67–73.

Zisman, S. (1990) *Mangrove Habitats of Belize. Extent, Characteristics and Research Needs. Report to the Natural Resources Institute of the Overseas Development Administration for the Tropical Forestry Action Plan*. London: Overseas Development Administration.

16

Ecological Implications of Wetland Conversion and Utilization around Metropolitan Lagos, Nigeria

Chebo K.A. Asangwe

With some 50 per cent of the global human population occupying coastal zones, it is hardly surprising that such areas frequently experience severe land pressure and a high risk of degradation. This is particularly the case with coastal wetlands. These diverse and resource-rich habitats are all too often the scene of encroachment, including urban settlement. In this chapter, the ecological implications of such settlement are considered in respect of metropolitan Lagos, Nigeria, where significant degradation of coastal wetland is currently occurring as a function of large-scale land reclamation. Consideration is given to the impact of the reclamation, and some longer-term policy implications are discussed.

Coastal Wetlands around Metropolitan Lagos

One of the locations of most pronounced environmental degradation in Nigeria today is its low-lying and swampy coastal zone. In this area, the relentless action of tides and ocean surges, allied to the flows of coastal rivers, has caused the inundation of low-lying areas around Lagos and created extensive wetlands. In spite of their poor drainage and infertile soils, such areas are densely vegetated and biologically diverse under natural conditions. Anthropic intervention, however, has significantly modified the environment, intensifying the natural processes of flooding and erosion. At present, an estimated 60 per cent of metropolitan Lagos, particularly in its western area, extends over what were originally natural wetlands (Asangwe, 1992). Much of the land in question has been reclaimed over the last thirty years as a result of modern hydraulic sand-filling techniques.

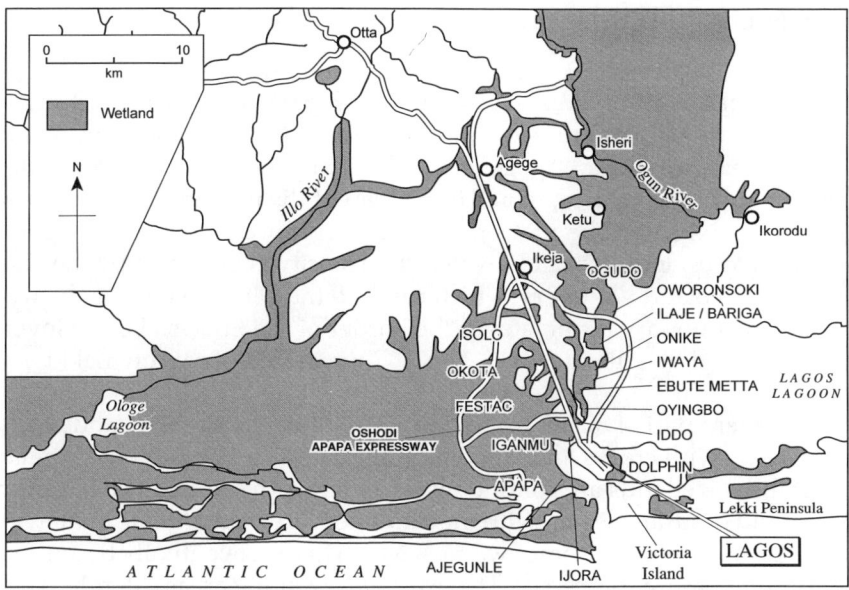

Figure 16.1 Location map of Lagos metropolitan area, Nigeria.

The early development of Lagos was largely controlled by the physical setting of Lagos lagoon, where shipping access was provided to the Atlantic Ocean (Figure 16.1). Initial reclamation of wetlands around the city was undertaken by the British colonial administration in the early twentieth century, mainly as a sanitary measure to control the threat of malaria. Reclamation was continued by later administrations and is generally recognized by local residents as a suitable means of providing the requisite dry land for urban expansion. The need for such land is considerable and continuing, given the rapid population growth and expanding economic activities of the metropolitan area. Thus, higher priority has generally been given to socio-economic than environmental factors in converting coastal wetlands to more 'productive' use. Of late, such conversion has mainly involved the construction of highways, port facilities, housing estates, industrial estates, power plants and the like. This is in accord with the Lagos State Regional Plan (1980–2000), which stresses the need for additional land to meet the demands of industrial, residential and other vital land uses (New Towns Development Authority, 1986).

The large-scale expansion of Lagos and its suburbs over recent decades has created various ecological problems in wetland areas. In spite of the physical constraint on urban expansion presented by the existence of Lagos lagoon on the eastern flank of the city, it is likely that expansion will continue across the wetlands to the west, unless planning constraints are applied.

Lagos Metropolitan Area

The Lagos metropolitan area lies in the central part of Lagos State. North of the city, the coastal plain rises to approximately 14 m above sea level, where the settlements of Ikeja, Agege and Isheri are located, but Lagos itself occupies a lower-lying area, which forms part of the main sedimentary basin of southwest Nigeria. The metropolitan area comprises a maze of lagoons, creeks, estuaries and associated wetlands. The city is bounded by the channels of the Owo and Illo rivers, which run into the Ologe lagoon to the west. The eastern boundary terminates at the border of the Eti–Osa Local Government Area and the Ibeju–Lekki Local Government Area on Lekki peninsula.

The southern part of the metropolitan area is characterized by sandy ridges that alternate with creeks and lagoons lying parallel to the present shoreline. Most of these wetlands are continually supplied with sediments from inland sources, and gradually in-fill with mud and decomposed organic matter. Old lagoons progressively become swamps, which in turn eventually become dry land through natural processes. The extensive swampy areas extending from Ebute Metta and Oyingbo, through Iganmu to Ologe lagoon exemplify these transitional stages (Figure 16.1). When such areas become dry land, they are suitable for settlement and are generally protected from tides and ocean surges.

The evolution of this depositional landscape has been decisively influenced by the humid climate, low relief, infertile soils and dense wetland vegetation. The latter, which largely comprises saline and freshwater swamp forest, is a resource-rich habitat, but it is also the scene of continuous human encroachment and destruction that currently cause significant environmental degradation.

The Lagos metropolitan area is easily the most urbanized area along the coast of West Africa. According to the provisional 1991 census figures, the area, covering 3577 km², has a total population of some five million people, while the remainder of Lagos State contains a further 500 000 people. The population of the metropolitan area is currently increasing very rapidly and is likely to reach between 12 and 15 million people by the year 2000. This will clearly induce further rapid physical expansion of the metropolis and ensure continuing massive alteration of the natural environment. In particular, it is likely to involve further large-scale encroachment on wetland areas on a scale unprecedented in West Africa.

Most of this encroachment will tend to occur west of the city, although significant wetland reclamation is currently taking place along the western margin of Lagos lagoon, which bounds the city to the east. The most extensive of the vulnerable wetlands stretches from the estuary of the Ogudu stream, through the Ogun river floodplain and estuary, to the estuary of the

Majidun river near Ikorodu. This densely settled zone, which extends to Oworonsoki in the south and as far as the Iwaya and Onike swamps, has already experienced extensive road and bridge development and attracted much new settlement. To the south of Oworonsoki, settlement has also extended across the marginal depressions of the Lagos lagoon in the vicinity of Ilaje and Bariga (Figure 16.1).

Wetland Ecosystems

Coastal wetlands are low-lying areas that constitute the interface between terrestrial and aquatic ecosystems. Their water tables are characteristically at, near or above the land surface for a significant period of the year. In the vicinity of metropolitan Lagos, the wetlands are essentially marine, lagoonal and estuarine, and their physical complexity creates an environment that is diverse and distinctive in biological terms. In many respects, the zone has a high natural resource potential, but it remains vulnerable to human interference and prone to degradation. The wetlands to the northeast typically have mangrove communities along the lagoon shoreline. Inland, on drier sites, bush thicket and grassland communities commonly exist which supply an increasing demand for fuelwood.

As a result of the relationship between the habitat and its natural vegetation, the coastal wetlands have traditionally acted as reservoirs for excess surface run-off and have served to limit coastal flooding and erosion. They have also acted as natural sieves that filter out and absorb sediments and organic matter brought by surface run-off to the coastal zone. They have thus protected water quality by removing pollutants, and also recharged aquifers in the coastal zone.

As elsewhere in the tropics, the role of mangroves in the vicinity of Lagos has been particularly important. Their presence in areas like the backshore zones of Victoria Island and Lekki peninsula has effectively controlled erosion along both marine and lagoonal shorelines. The mangroves trap and bind fine alluvium around their roots, and serve as a buffer against storm surges from waves and tidal currents. By acting as reservoirs for flood waters, the mangrove areas further dissipate storm energy and provide natural flood control. Conversely, where there has been a significant human impact on the mangroves, their status and natural functions have inevitably been damaged.

Wetland Reclamation and Degradation

Given the rapid population growth and associated land pressure in the Lagos area, the presence of extensive wetlands has often been seen as a constraint on urban development. Wetlands are thus commonly perceived as waste-lands and as hazards to health that should be in-filled and used more productively. As a result, they have lately been subject to an unprecedented rate of reclamation, and, as a result of the intense competition for land, have experienced haphazard encroachment of multiple land uses. These uses include housing estates, industrial developments including utilities like power, water and sewage, excavation for construction materials, road build-ing, drainage works, and even agricultural activities.

The main process of wetland reclamation around metropolitan Lagos involves the displacement of surface water by hydraulic sand filling. Such filling results in a total loss of habitat and of associated vegetation and fauna in the affected areas, as well as modifying infiltration capacities and run-off rates. Public reclamation projects near Victoria Island and along the Oshodi–Apapa expressway are notable examples of these sand-filling operations. Other areas where sand filling is occurring include Dolphin estate.

As well as public reclamation schemes, small-scale private reclamation of swamp and mudflat areas also occurs, using admixtures of sand, clay, domestic waste and other organic and inorganic residues. This is common in suburban areas like Bariga, Onike, Ajegunle and Orile Iganmu. It is estimated that more than one million people in metropolitan Lagos are involved in small-scale, private reclamation schemes.

Where both public and private land reclamation has been undertaken, problems commonly arise as a result of poor reclamation practices. This is particularly the case along the margins of lagoon inlets, creeks and tidal mudflats, where domestic refuse, industrial and other solid wastes, and poorly sorted sands have been used as in-fill. In such areas, frequent collapse of building structures occurs, as in the vicinity of Bariga and Orile Iganmu. The main problem is that inadequate drying out and compaction of in-fill materials is allowed to occur before building construction commences. Given the general scarcity of land in the metropolitan area, much building has lately been multi-storey, which increases the likelihood of collapse. In addition, widespread ground subsidence has occurred, particularly on poorly re-claimed lagoon margins. It is estimated, for example, that over 80 per cent of residential buildings in the Onike–Iwaya and Ilaje–Oworonsoki areas of Lagos have experienced tilting and subsidence (Asangwe, 1992).

A further problem apparent in reclaimed wetlands is a continuing suscept-ibility to perennial flooding. Thus, as a result of increasing reclamation, the overall ability of remaining wetland areas to act as reservoirs for surface water run-off is reduced, and local flooding continues to affect areas that are supposedly reclaimed. This is the case in peripheral areas of the city, such as Bariga, Iganmu and Victoria Island, where serious flooding may occur even after moderate rainfall. This remains one of the most serious environmental hazards affecting metropolitan Lagos today, and, notably in the case of the Victoria Island annexe, has necessitated huge investment in mitigation measures dating back to 1958 (Asangwe, 1993). A contract of ₦ 700 million (*c.* £9 million at 1994 rates) has recently been concluded for pumping sand on to the Victoria Island beach, while such measures as placing sand bags on the shoreline are also being undertaken.

As well as the problems encountered in reclaimed wetlands themselves, serious degradation also occurs in adjacent, unreclaimed wetlands. This is evident, for example, where roads and bridges have been constructed across larger wetlands, which are consequently fragmented. A major bridge across the margins of the Lagos lagoon linking Adekunle to Oworonsoki, is routed over the Onike, Ilaje–Bariga, Oworonsoki and Ogudo wetlands, fragmenting the latter. Likewise, the Oshodi–Apapa expressway has isolated several wetlands. The Kirikiri wetland has been cut off from the Coker and Festac wetlands and the Okota and Isolo wetlands have been separated from the Itire and Ijeshatedo swamps. Such isolation of wetlands reduces or precludes regular inputs of water and sediments. This disrupts the natural processes of sedimentation and reclamation, and the seriousness of flooding is often increased in cut-off areas.

Even where flows of water into unreclaimed wetlands are sustained, the water quality is often impaired. This is the case along the Oshodi–Apapa expressway and the Lagos–Badagry highway, where housing schemes and industrial developments have been established. The present network of rivulets and creeks in these areas now mainly functions as drains for reclaimed land and carry substantial quantities of domestic and industrial wastes. Elsewhere in Lagos, the development of chemical and other industries has led to major pollution inflows into unreclaimed wetland areas, causing serious damage to fish stocks and other wildlife. This is particularly the case in the vicinity of the Iganmu and Oregun–Ikeja industrial estates, where fish catches in the adjoining wetlands have been drastically reduced in recent years. A similar impact is evident in the Iddo, Ijora, Orile Iganmu and Bariga areas, where domestic sewage from housing estates has polluted adjacent, unreclaimed wetlands. In addition, there has been serious contamination of waters flowing into the Lagos lagoon, which contains valuable fish stocks, has considerable aquacultural potential, and provides an important water supply for the inhabitants of Lagos. Such pollution has lately attracted the attention of the Federal Environmental Protection Agency, which has prosecuted and fined several commercial companies for water pollution, as well as requiring them to obtain equipment to control pollution levels in future.

Conclusion

The development of industrial and commercial activities within metropolitan Lagos, and hence its current expansion, are largely due to its position on Lagos lagoon with its shipping access to the Atlantic Ocean. The margins of the lagoon complex have long been a major focus of wetland reclamation and urban expansion. Competition for use of this reclaimed land has been considerable, making it very valuable, but also vulnerable. To the west of Lagos, significant reclamation of wetland has also occurred. In general, this land reclamation, whether for industrial, commercial or residential purposes, is of a permanent nature, and no possibility exists for restoring the original wetlands. Much damage is also being inflicted on unreclaimed wetland around metropolitan Lagos, as a result of both pollution and fragmentation. Overall, there is clear evidence of widespread degradation of wetlands in the area.

Given the expected rate of growth of metropolitan Lagos in the future, pressure to reclaim remaining wetlands in the vicinity of the city is likely to continue. In these circumstances, it is clear that regular environmental monitoring as a basis for land management, including conservation planning, is required for the whole area. In particular, it is suggested that, as a means of conserving at least some valuable wetland habitats in the vicinity, an official policy should be established to promote additional northward growth of the metropolis over the undulating and free-draining coastal plain around Ikeja and Agege and towards Otta in Ogun State. Another axis for urban growth should be promoted from Isheri towards Ogere and Sagamu in Ogun State. In this respect, it is important to conserve the wetlands separating Ketu and Ikorodu as this area is the prime source of fresh water, fish and sediments for the Lagos lagoon. Encroachment thereon would have very damaging ecological implications for the entire Lagos metropolitan area, as also would continuing loss of the coastal wetlands.

References

Asangwe, C.K.A. (1992) Aspects of geomorphology and coastal environmental management of metropolitan Lagos. Unpublished paper.

Asangwe, C.K.A. (1993) Issues and problems in controlling the threats of erosion along the Nigerian coastline. In O. Magoun (ed.), *Coastlines of Western Africa*. New York: American Society of Civil Engineers, Vol. 5, pp. 309–24.

New Towns Development Authority (1986) *Lagos State Regional Plan (1980–2000)*. Lagos: Lagos State Military Government.

PART V

Urban and Industrial Degradation in the Tropics

17

Environmental Impacts of Urban–Industrial Development in the Tropics: An Overview

Robert B. Potter

Over the last half century, the world has experienced its fastest ever rate of urbanization. On the ground, the era has witnessed the growth of very large cities and the associated challenges of housing, jobs and services. This has occurred especially in poorer countries – the nations that are often referred to as the Third World, many of which are coterminous with the tropical world. This represents a change from the past in that rapid urbanization is no longer directly associated with the rich, industrial countries of the temperate world. Indeed, in many rich countries, central cities are losing populations due to inner-city decay and counter-urbanization (Gilbert and Gugler, 1992; Potter, 1993a).

During the period 1950 to 2025, the level of world urbanization will have more than doubled, from 29 to 61 per cent. Between 1960 and 1980, the world's urban population grew by just under 17 per cent per decade. By the year 2000, 46.7 per cent of the world's population will be urban, made up of 74.8 per cent of the population of More Developed regions and 39.5 per cent of the population of Less Developed regions. During the period 1950 to 2025, the absolute number of urban dwellers in Less Developed regions will have increased by a staggering fourteen times, from 300 million to 4 billion. By the end of the millennium there will be two city dwellers in the Less Developed regions for every one in the More Developed regions. Thus, United Nations' statistics indicate that of a world urban population of some 3090 million in the year 2000, 2080 million will be found in Less Developed regions and 1010 million in More Developed regions. These data relate to a projected world population of 6 billion. The salient point is that by the year 2025, the ratio will have increased from 2:1 to 3:1 (United Nations, 1989; Potter, 1994).

The growth of urbanization in the tropical world is shown clearly by the analyses of 'million cities' that have periodically been carried out by geographers (e.g. Mountjoy, 1968, 1976). Between the 1920s and 1980s, there was a shift in the average latitude of cities with more than one million inhabitants from 44° 30' to 34° 07'. United Nations data indicate that by the year 2000, the three largest cities in the world will be Mexico City, São Paulo and Shanghai, each with a population in excess of 23 million. This contrasts starkly with the 1950s, when the three largest conurbations were New York, London and the Rhine–Ruhr (Table 17.1).

Table 17.1 The largest cities in the world, 1950 and 2000.

	1950			*2000*	
Rank	*City*	*Population (million)*	*Rank*	*City*	*Population (million)*
1	New York	12.3	1	Mexico City	31.0
2	London	10.4	2	São Paulo	25.8
3	Rhine-Ruhr	6.9	3	Shanghai	23.7
4	Tokyo	6.7	4	Tokyo	23.7
5	Shanghai	5.8	5	New York	22.4
6	Paris	5.5	6	Beijing	20.9
7	Buenos Aires	5.3	7	Rio de Janeiro	19.0
8	Chicago	4.9	8	Bombay	16.8
9	Moscow	4.8	9	Calcutta	16.4
10	Calcutta	4.6	10	Jakarta	15.7
11	Los Angeles	4.0	11	Los Angeles	13.9
12	Osaka	3.8	12	Seoul	13.7
13	Milan	3.6	13	Cairo	12.9
14	Bombay	3.0	14	Madras	12.7
15	Mexico City	3.0	15	Buenos Aires	12.1

Source: after United Nations (1989).

Third World countries have also been industrializing over the past twenty years. However, this process has not only been far slower than that of urbanization, but has also been more localized. During the colonial period, Third World countries, as producers of primary products, were never encouraged to develop industry. Between 1960 and 1983, the relative share of manufacturing to Gross Domestic Product in the Third World increased from 15.6 to 17.5 per cent and was predicated on the so-called 'New International Division of Labour'. However, Less Developed countries still only contributed 11.9 per cent to total world manufacturing output in 1985, having increased from 10.3 per cent in 1975 (Chandra, 1992). The increase has been concentrated in a very few Newly Industrializing countries, particularly those that the World Bank refers to as the 'Asian Tigers'. The leading exporters of manufactures in 1985 were South Korea, which accounted for 22.3 per cent of the total manufactured exports of all Less Developed countries, Hong Kong (16.7 per cent), Mexico (9.6 per cent), Brazil (9.0 per cent) and Singapore (7.1 per cent). Thus, 64.7 per cent of the manufacturing exports of the entire Third World was accounted for by a mere five nations.

The recent period of the New International Division of Labour is characterized by a trend towards *global divergence* whereby, through the activities of multinational companies in particular, patterns of ownership, capital accumulation and production are becoming more concentrated and diverse at the global level (Armstrong and McGee, 1985; Potter, 1993b; Potter and Dann, 1994). The severe competition means that the process of industrialization is neither monitored nor managed very closely. Consequently, low-income nations are prepared to ignore health, safety and environmental issues in their search for investment and employment. The best example is the recent involvement of Third World countries in the re-processing of highly toxic chemicals exported from developed nations. The trade is prompted by the lower costs and lack of regulations in many Third World countries. Clapp (1994) reports that, in the late 1980s, of the 30 to 45 million t of hazardous waste that crossed international borders, 20 per cent went to Third World nations.

In this introductory chapter, some environmental ramifications of urbanization and industrialization in the tropical world are considered. A pivotal theme is that the very different character of the two processes is itself a critical and pressing issue. Two avenues are pursued: the first deals with the relations between urban development and environmental circumstances, and the second explores the links between industrialization and urban environmental conditions and change.

Urbanization and Environment

Towns and cities can provide healthy and efficacious environments for their inhabitants. The function of towns and cities is to centralize facilities, services and opportunities, both social and economic. Cities offer job opportunities in both the formal and informal sectors, and, where jobs exist, average wage rates increase progressively with city size (Hoch, 1972). In theory at least, it should be possible to provide housing, water, sanitation, paved roads, transport systems and recreational and open spaces at lower unit cost than in rural areas.

But all too frequently, population densities are high, resources are insufficient, structural poverty is rife, and the needs of city dwellers are not being met. For instance, it is estimated that more than 50 per cent of Third World urban residents live in sub-standard dwellings, and this has a direct bearing on their environmental circumstances (Potter, 1992), a theme that is developed by Williams (Chapter 19) and Lloyd-Evans and Potter (Chapter 20). In 1987, the World Bank estimated that less than 60 per cent of the urban population in Developing Countries had access to adequate sanitation and only 33 per cent were connected to a sewer system. It is also evident that 20 to 30 per cent of urban dwellers regularly purchase supplies of drinking water from mobile vendors (McAuslan, 1985; Hardoy *et al.*, 1992). Some sense of the magnitude of the problems can be gained from a single city. With 6 million inhabitants, Hong Kong produces 23 300 t of solid waste, 21 t of floating refuse, 2 million t of sewage and industrial waste per day, along with 100 000 t of chemical waste per year (Chan, 1994). Major difficulties are now being experienced in finding sufficient landfill sites. Other problems stem from noise pollution from vehicles, aircraft and construction. And, as Chan (1994) observes, it is the poor of Hong Kong who suffer the most as a result of the environmental degradation. Although the best health facilities available in Third World cities are frequently as good as those found in Western cities, pathogens in the environment pose serious threats. The main scourges are diarrhoea, dysentery, typhoid and cholera, which disproportionately affect the poor. Overcrowding in residential areas means that diseases like tuberculosis, influenza and meningitis, once contracted, are easily transmitted (Hardoy *et al.*, 1992).

Rapid urban growth has resulted in millions of people being forced to build houses on hazardous sites, such as steep hillslopes. This is as true of small urban areas like Kingstown, the capital of St Vincent and the Grenadines, as it is of well-known cities like Rio de Janeiro in Brazil, Medellín in Colombia, and Caracas in Venezuela. However, the outcome is more frequently reported by the media in the case of large cities. Jimenez-Diaz (1994) found that, by 1985, 61 per cent of the total population of metropolitan Caracas lived in *barrios*, and that 67 per cent of the land occupied by such settlements was unstable enough to justify eviction. During the 1950s and 1960s, landslides in Caracas averaged about one per year. However, this increased dramatically to a mean of 25 per year between 1971 and 1979, and 33 per year between 1980 and 1987. From 1970 onwards, slope failures were mainly associated with the incidence of heavy rainfall, rather than earthquakes, as had previously been the case (Jimenez-Diaz, 1994). On 8 August 1993, for example, at least 150 people were killed and thousands left homeless when Tropical Storm Bret caused mudslides that affected hilly *barrios* in Caracas. The threat of flows and landslides is frequently aggravated by human action, notably the removal of the vegetation cover. This increases surface water run-off, which commonly includes sewage and other waste water. Elsewhere, low-income populations are forced to inhabit reclaimed or swampy areas or sandy beaches. Examples of the latter occur at Byera, Mangrove and Georgetown on the windward coast of St Vincent and the Grenadines (Potter, 1995).

Urbanization frequently increases the demand for non-indigenous goods, and this may cause environmental degradation either directly or indirectly. For example, high levels of migration overseas, whilst generating remittances, may result in idle land and a general disinclination to engage in agriculture (Thomas-Hope, 1992). Some of the related issues involved in promoting agricultural intensification within the urban area are considered by Bowyer-Bower (Chapter 18). There is the need in all settings to look closely at the balance that exists between indigenous and foreign influences and modes of development. An example is the direct link that exists between tourism and child prostitution in many Third World countries, as in the case of Bangkok, Thailand. Again it is apparent that social and environmental degradation are inextricably linked.

Industrialization and the Environment

Industrial concentration can maximize economies of scale for productive enterprises. It can also reduce the costs of enforcing regulations governing environmental–occupational health standards and pollution control in urban and industrial zones. In the mining and extractive industry, concentration and localization often reflect the nature of the reserves. McShane (Chapter 21) deals specifically with the potential impacts of a copper mine. In this context it is worth noting that the contribution of Third World cities to global environmental problems remains relatively small – it is mainly cities of the Developed World that are emitting greenhouse gases. Thus, the main threat from industry located in the Third World is to the local population. Even so, Third World cities are threatened by global warming, which may lead to more severe storms as well as sea level rise.

The contribution of the informal petty-commodity production and retailing sectors raises major issues. On the one hand, the informal sector of the economy provides jobs and homes, often where the state cannot or will not do so. On the other hand, by definition, the growth and development of the informal sector are unregulated. The outcome is likely to be more garbage, poor environmental upkeep, and visual blight, as reported in Nigerian cities (Olu Sule, 1981; Ebong, 1983). It is estimated that perhaps 70 per cent of the workforce in Third World countries are employed in the informal sector (Santos, 1979), and the proportion has increased with the global recession and structural adjustment programmes. This theme is explored by Lloyd-Evans and Potter (Chapter 20) in the southern Caribbean.

A recent extreme event exemplifying the issue of informality and urban environmental degradation is the execution of street children in Rio de Janeiro and São Paulo, Brazil. This prompted an Amnesty International advertisement which shows a photograph of a Brazilian policeman pointing a gun at a homeless child in São Paulo, under the deliberately ambiguous and provocative caption, 'Everyday, children are murdered in Brazil. The police are doing all they can to help.' The accompanying text relates the events of 23

July 1993, on which day it is reported that eight children were murdered by hooded policemen. It also reports that 'death squads' killed 320 children in Rio de Janeiro in the first half of 1993. This extreme human rights issue shows that structural poverty is seen as a form of urban environmental despoliation which is to be 'cleaned up', even if it means executions. It is ironic this should have occurred immediately prior to the United Nations Conference on Environment and Development in Rio de Janeiro. It shows how easy and convenient it is for administrations to confuse the symptoms with the causes of a particular disease. It also exemplifies a general theme of this volume, that environmental degradation often exists cheek by jowl with social degradation and that the two are inextricably linked.

The rising tide of informal-sector homework carries with it many problems, especially for women and children in respect of health and safety, for once again regulation and protection are usually minimal. Risks in the work environment, whether at home or in the factory, include dangerous concentrations of toxic chemicals and dust, carcinogens, inadequate protection from machinery, and fires and explosions. Hardoy *et al.* (1992) cite 32.7 million occupational injuries per year globally, with as many as 146 000 deaths. Activities such as processing cement, asbestos, pesticides, plastics, iron and steel, textiles and leather are particularly hazardous. Frequent environmentally related occupational diseases include silicosis, emphysema, tuberculosis and hearing loss. Such diseases result from the exploitation of a workforce as cheap labour, and one which happens to provide its own housing.

In the area of political economy and dependence, Third World countries are lacking in power and must accept the conditions offered by multinationals and other large companies. Nanton (1983) cites the case of St Vincent and the Grenadines where, despite the existence of an official minimum wage, a foreign multinational demanded and got permission to pay workers at what it euphemistically described as a 'training rate'. Jobs at any price may involve cutting costs, along with poor safety standards. It is tempting to view these situations in terms of articulation theory, both with regard to housing and employment. Articulation theory focuses on modes of production under capitalism, arguing that pre-capitalist forms are preserved when this runs in favour of the system, but are swiftly dissolved in other cases (McGee, 1979; Potter, 1990).

The Bhopal tragedy is the extreme example of what can happen environmentally if industrial development proceeds on the basis of maximum return and insufficient monitoring and regulation. On 3 December 1984, a disastrous explosion occurred in a storage tank containing methyl isocyanate gas at the Union Carbide plant in Bhopal, India. Realistic estimates of the human death toll run at around 3000, although the official count is 1754 (Shrivastava, 1992). The Bhopal plant was an unprofitable operation which was 'ignored by the top Union Carbide officials' (Shrivastava, 1992). This, combined with the unregulated development of two large slum colonies across the road from the plant and housing several thousand people, set the scene for a major disaster. A report marking the tenth anniversary of the disaster highlighted the continuing suffering of many who survived, with victims facing breathlessness, dry coughs and chronic lung diseases (*Guardian*, 1994).

Another example of the human and environmental consequences of unregulated industrial development is provided by a spate of fires which have claimed the lives of workers, particularly women, in Third World factories. A newspaper report of 12 May 1993 chronicled a case where a raging fire killed at least 200 workers, and injured several hundred more, in a toy factory in an industrial zone on the outskirts of Bangkok, Thailand (*Guardian*, 1993a). The fire apparently ripped through the four-storey building so fast that few of the 1800 workforce were able to reach the fire escapes. There were no smoke alarms, and it was reported that two previous fires had occurred in the months immediately before the incident. Yet the factory bosses had not substantially altered working conditions nor taken any extra precautions. The report commented that the incident 'exposed on a horrific scale what is believed to be a widespread failure in many factories and workshops to observe even rudimentary precautions against fire' (*Guardian*, 1993a). The incident was swiftly followed by a number of factory fires in China's special development zones (Gittings, 1993).

These tragedies should be set against the latest pronouncements of the World Bank and the United Nations Development Programme which stress the importance of urban productivity (United Nations, 1992; World Bank, 1992). But as the Thai example shows, the social cost of such productivity is immensely high. A leader in *The Guardian* (1993b) warned that 'The World Bank and related international economic bodies need to look harder at the economic realities behind the productivity figures of Third World countries.' For how long will a passive labour force accept appalling conditions for the sake of a regular wage? The environmental and social consequences of such productivity have to be seriously questioned. In any case, the remaining Less Developed countries cannot now industrialize in the same way that the Newly Industrializing countries have (Griffiths, 1991). The World Bank and the United Nations Development Programme currently stress the role of the state as enabler and facilitator, but this will scarcely produce the environmental and social monitoring that is so pressingly needed in urban contexts.

Conclusion

In spite of the latest World Bank pronouncements, there is one issue over which we are in agreement. The 1991 World Bank report observes that 'more attention should be devoted to reversing the deterioration of the urban environment' (World Bank, 1992, p. 200). In addition, in overall terms, there has been all too little urban research on the Third World in the last decade. It is hoped that these shortcomings will be corrected in the 1990s and beyond, and that particular attention will be devoted to the topic of change and environmental degradation in fast-growing towns and cities, whether they be small, large or intermediate in scale. The substantive chapters in this section of the volume serve this objective.

References

Armstrong, W. and McGee, T.G. (1985) *Theatres of Accumulation: Studies in Asian and Latin American Urbanization*. London: Methuen.

Chan, C. (1994) Responses of low-income communities to environmental challenges in Hong Kong. In H. Main and S.W. Williams (eds), *Environment and Housing in Third World Cities*. Chichester: John Wiley and Sons, pp. 133–50.

Chandra, R. (1992) *Industrialisation and Development in the Third World*. London: Routledge.

Clapp, J. (1994) The toxic waste trade within less-industrialised countries: economic linkages and political alliances. *Third World Quarterly*, **15**, 505–18.

Ebong, M.O. (1983) The perception of residential quality: a case study of Calabar, Nigeria. *Third World Planning Review*, **5**, 273–85.

Gilbert, A. and Gugler, J. (1992) *Cities, Poverty and Development*. Oxford: Oxford University Press.

Gittings, J. (1993) Sixty die as blaze engulfs Chinese factory. *Guardian*, 21 November.

Griffiths, W.H. (1991) The applicability of the East Asian Experience to Caribbean countries. In Y.-k. Wen and J. Sengupta (eds), *Increasing the Competitiveness of Exports from Caribbean Countries*. Washington DC: World Bank, pp. 91–100.

Guardian (1993a) Toy factory folds like house of cards in Bangkok inferno, 12 May.

Guardian (1993b) Leader comment, 12 May.

Guardian (1994) Out of sight and out of mind, 14 March.

Hardoy, J.E., Mitlin, D. and Satterthwaite, D. (1992) *Environmental Problems in Third World Cities*. London: Earthscan Publications.

Hoch, I. (1972) Income and city size. *Urban Studies*, **9**, 299–328.

Jimenez-Diaz, V. (1994) The incidence and causes of slope failures in the barrios of Caracas, Venezuela. In H. Main and S.W. Williams (eds), *Environment and Housing in Third World Cities*. Chichester: John Wiley and Sons, pp. 125–50.

McAuslan, P. (1985) *Urban Land and Shelter for the Poor*. London: Earthscan Publications.

McGee, T.G. (1979) Conservation and dissolution in the Third World city: the shanty town as an element of conservation. *Development and Change*, **10**, 1–22.

Mountjoy, A. (1968) Million cities: urbanization in developing countries. *Geography*, **53**, 365–74.

Mountjoy, A. (1976) Urbanization, the squatter and development in the Third World. *Tijdschrift voor Economische en Sociale Geografie*, **67**, 130–7.

Nanton, P. (1983) The changing pattern of state control in St Vincent and the Grenadines. In F. Ambursely and R. Cohen (eds), *Crisis in the Caribbean*. London: Heinemann, pp. 223–46.

Olu Sule, R.A. (1981) Environmental pollution in an urban centre: waste disposal in Calabar. *Third World Planning Review*, **3**, 419–31.

Potter, R.B. (1990) Cities, convergence, divergence and Third World development. In R.B. Potter and A.T. Salau (eds), *Cities and Development in the Third World*. London: Mansell, pp. 1–11.

Potter, R.B. (1992) *The Quality of Housing in Barbados: A Geographical Analysis*. Mona: Institute of Social and Economic Research, University of the West Indies.

Potter, R.B. (1993a) *Urbanisation in the Third World*. Oxford: Oxford University Press.

Potter, R.B. (1993b) Urbanisation in the Caribbean and trends of global convergence–divergence. *Geographical Journal*, **159**, 1–21.

Potter, R.B. (1994) Urbanisation and development in the Third World. In P.P. Courtney (ed.), *Geography and Development*. Melbourne: Longman Cheshire, pp. 89–116.

Potter, R.B. (1995) *Low-Income Housing and the State in the Eastern Caribbean*. Mona: The Press University of the West Indies, Barbados, Jamaica, Trinidad and Tobago.

Potter, R.B. and Dann, G.M.S. (1994) Tourism and postmodernity in a Caribbean setting. *Cahiers du Tourisme*, series C, 185, 1–45.

Santos, M. (1979) *The Shared Space: The Two Circuits of the Urban Economy in Underdeveloped Countries*. London: Methuen.

Shrivastava, P. (1992) *Bhopal: Anatomy of a Crisis*. London: Paul Chapman.

Thomas-Hope, E.M. (1992) *Explanation in Caribbean Migration: Perception and Image: Jamaica, Barbados, St Vincent*. Basingstoke: Macmillan Caribbean.

United Nations (1989) *Prospects for World Urbanisation 1988*. New York: United Nations Department of International Economic and Social Affairs.

United Nations (1992) Cities, people and poverty: urban development co-operation for the 1990s. In N. Harris (ed.), *Cities in the 1990s: The Challenge for Developing Countries*. London: UCL Press, pp. 212–18.

World Bank (1992) Urban policy and economic development: an agenda for the 1990s. In N. Harris (ed.), *Cities in the 1990s: The Challenge for Developing Countries*. London: UCL Press, pp. 199–211.

18

Environmental Protection and the Control of Urban Agriculture in Harare, Zimbabwe

Tanya A.S. Bowyer-Bower

On visiting Harare, the capital city of Zimbabwe, it is not unusual to notice vegetable gardens on tracts of public land within the residential and industrial sectors. From pre-independence days, however, such cultivation has been discouraged, supposedly for reasons of environmental protection. Yet, no evidence of environmental degradation from such cultivation has been documented, nor has the need for the cultivation been researched. During 1992, a preliminary investigation of the topic was undertaken. The information included in the chapter is derived from field survey, archive research of city council minutes and newspaper reports, and interviews with city officials. The conflict between the apparent need for the cultivation of public land and for its control is reviewed, and current policies and possible planning responses are assessed.

Cultivation Practices

The cultivation of public land within Harare is a long-established practice. It was first studied by Mazambani (1982), who, by aerial photograph survey, determined an increase in its practice from 267 ha to 4762 ha between 1955 and 1980. Newspaper reports since 1980 suggest that the area is still increasing (*The Herald*, 16/12/84, 28/11/90, 1/3/92). In 1992, the main crop encountered was maize, with some sweet potatoes. Other crops included rice, squash, watermelon, groundnuts and peas. Cultivation is by hand tilling, which is largely carried out by women and children, although men join them at weekends.

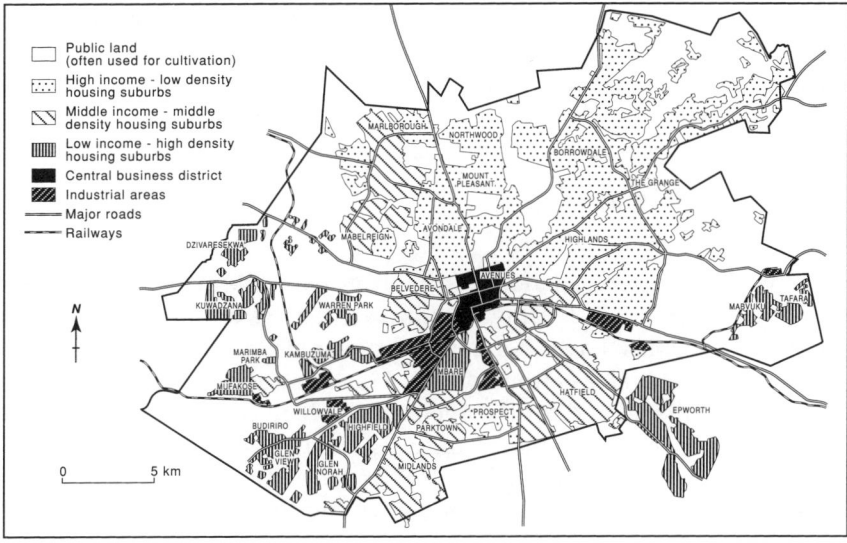

Figure 18.1 Sectoral structure of the city of Harare, Zimbabwe, showing large tracts of public land often used for cultivation.

In 1992, the cultivation was most widespread on vacant land in the lower-income suburbs of Harare, which are largely concentrated in the southwest, including areas like Budiriro, Glen Norah, Glenview, Highfield, Kambuzuma, Warren Park, Mafakose, Marimba Park, Kuwadzana and Dzivaresekwa, and also in eastern areas like Mabvuku and Tafara (Figure 18.1). Cultivation also occurs on vacant residential plots in some middle- and higher-income areas, like Park Town, The Grange and The Avenues. Cultivation was also observed on road-side verges in industrial areas like Willowvale, along roads leading to lower-income suburbs, on land outside the boundaries of residential properties, and alongside railway lines. Less trampled land around bus stations, as at Kuwadzana, is also occasionally cultivated.

In addition to the above, the most extensive areas of municipal land used for cultivation are the *vleis*, which are seasonally waterlogged depressions in the headwater areas of drainage basins. The natural drainage from Harare is mostly radial from the city centre (Thompson, 1972). The *vleis* are extensive, but mostly unsuitable for building because of their seasonal waterlogging and the swelling clay soils that create problems for building foundations. Similar land has long been favoured for cultivation in rural areas. This is partly because the *vleis* are the first areas to collect and hold water at the onset of rains, thus permitting an early crop. Also they retain water well into the following dry season, which favours late cropping. *Vleis* are best cultivated by hand tilling, since the soils are susceptible to severe damage through use of heavy machinery or trampling by livestock. Various studies have extolled the virtues of the *vleis* for small-scale gardening such as that undertaken on public land in Harare (Rattray *et al.*, 1953; Theisen, 1975; Whitlow, 1983).

Control of Cultivation

The cultivation of public lands within Harare is illegal and newspaper articles describing the destruction of cultivated crops on public land by the municipal authorities have frequently appeared in the ten-year period 1982–92 (*vide The Herald*, various dates). However, the reasons for making such cultivation illegal and for the attempts made to control it are less clear-cut. During interviews, municipal officials cited reasons of environmental protection. These included the need to reduce soil erosion which may otherwise threaten municipal water supplies and to eliminate breeding sites for malarial mosquitoes.

The legislation restricting the cultivation of public land is twofold. Firstly, there are the streambank protection regulations of the Natural Resources Act (1952), which forbid agriculture within 30 m of any watercourse or on wetlands. Secondly, there are the land protection bye-laws of the Municipal Act (1973), which forbid cultivation in the city without the written approval of the authorities, and which permit the destruction of crops grown without such approval.

The streambank protection regulations target cultivation on the grounds that it involves removal of the natural vegetation and exposes the soil to erosion by wind and rain. Erosion rates are further increased by tillage which damages the soil structure. When cultivation is adjacent to a watercourse, the loose soil is more easily washed away and eroded, causing channel siltation, which, if extreme, prevents any surface water flow. The rationale for land protection in the Municipal Act is to control aspects of health, safety, amenity, convenience and general welfare within the city, including control of activities that are unsightly.

This legislation has not been without its critics. In rural communal areas, land in the vicinity of *vleis* is highly valued as a reliable and accessible source of water for humans and livestock, as productive grazing for livestock, and as well-watered land for garden cultivation (Rattray *et al.*, 1953; Cormack, 1972; Elwell and Davey, 1972; Grant, 1974; Theisen, 1975; Windram, 1983; Bell *et al.*, 1987). Fears that environmental degradation would result from such functions have been investigated and, of the functions in question, livestock production has been determined to be the worst offender (Rattray *et al.*, 1953; Elwell and Davey, 1972; Elwell, 1983). The value of the streambank protection regulations, which only limit cultivation activities, is therefore questioned (Elwell and Davey, 1972; Bell *et al.*, 1987; Whitlow, 1989; Scoones and Cousins, 1991). Meanwhile, the regulations stand without amendment.

The legislation has been criticized on other grounds as well. It has been seen as an interventionist strategy that was originally enforced by the colonial government for the furtherance of imperial interests rather than for the good of the country and its people (Moyo *et al.*, 1991; Scoones and Cousins, 1991; Gore *et al.*, 1992). In addition, there has been criticism of the government bodies responsible for environmental management and for the research, monitoring and implementation of the legislation, on the grounds that they are too diverse and fragmented (Ministry of Environment and Tourism, 1993). Confusion has arisen over the aims and motives of the legislation, and there have been inconsistencies and contradictions in its implementation (Moyo *et al.*, 1991; Gore *et al.*, 1992). This has further fuelled the argument about the interventionist nature of the legislation, although the argument is now directed at a 'petit-bourgeois' black elite rather than the colonial power.

The complexity of the attempted control of cultivation on public land in Harare illustrates the general problem. The hierarchy of organizations and bureaucracy involved in the control is not easily determined, but it includes the following: the Department of Health, Housing and Community Services responsible for evaluating and managing community needs; the Department of Town Planning and Works which is directly involved in land use planning; the Finance and Development Committee responsible for collecting rents from municipal land users; the municipal police who supervise use of the land; the Natural Resources Board which provides information on various aspects of the Natural Resources Act; the Department of Agricultural, Technical and Extension Services which carries out research related to land conservation and management; the Ministry of Education responsible for providing instruction to schoolchildren on the restrictions relating to urban cultivation; the Director of Works whose office controls cultivation in high-density suburbs; and the Town Clerk whose office is responsible for coordinating aspects of the above.

Various attempts have been made to enforce the legislation prohibiting the cultivation of public land in Harare. They include the following:

1   The defence of many of the cultivators has been that they are unaware of the illegality of their actions, and so the Department of Health, Housing and Community Services has undertaken air-drops of information leaflets. These state: 'In terms of the Natural Resources (Protection) Regulations 1975, you are hereby warned that no person shall cultivate within thirty metres of the naturally defined banks of a Public Stream. Should you defy this regulation, the appropriate authorities are entitled to destroy such illegal cultivation.' The leaflets have also been delivered

to households in high-density suburbs, and similar warnings have been advertised in local newspapers.

2   It is often difficult to identify seasonal watercourses on the ground, and so the Department of Town Planning and Works has physically demarcated the restricted areas with wooden stakes with notices attached, stating (in English and Shona): 'This notice is 30 m from the river bank. Cultivation between this notice and the river bank is illegal.' It is reported that Z$25 000 (£2780, at 1994 rates) has been spent on these notices, but unfortunately many of the stakes have been uprooted for homestead fencing. Iron supports firmed with concrete were recommended for future use (*The Herald*, 6/9/85).

3   The municipal police have assumed responsibility for patrolling wetlands and watercourses, for reporting illegal activities, and for issuing warnings to those involved (*The Herald*, 23/10/82). However, patrols have been limited due to a shortage of fuel.

4   The City Council has employed 'slasher squads' to destroy crops grown in contravention of the legislation, i.e. being within 30 m of a watercourse or being grown without permission.

5   Other actions to inform residents of the legislation have included broadcasts on National Radio, posters distributed to primary and secondary schools, and a taped message from the Mayor that has been relayed by loudspeaker in high-density suburbs.

Considerable efforts have thus been made to control cultivation on public land in the name of environmental protection. Further to this, in July 1988, the Finance and Development Committee requested the Town Clerk to report on all activities on municipal land for which no rents were being received. Activities ranged from open-air prayer meetings to the cultivation of public land. This led to a further spate of crop slashing by the authorities, and more anger among those whose crops were destroyed. In places, the destruction extended more than 30 m from watercourses; whether this was for reasons of environmental protection or simply an enforcement of legislation for raising additional government revenues is unclear. In spite of attempts at control, cultivation on public land is increasing, and recurrent conflict occurs between the cultivators and the municipal authorities (*The Herald*, 29/12/84, 29/12/91, 1/3/92).

The Need for Cultivation

Investigation of the reasons for continued cultivation of public land in the face of determined opposition from the authorities is a prerequisite to formulating more effective management policies. In Mazambani's (1982) view, the cultivators are the urban poor who do not earn enough to purchase all the food they need. Cultivation of any available land was thus seen as a symptom of increasing economic hardship. Since the cultivation is largely associated with lower-income, higher-density areas of the city, Mazambani's interpretation would appear correct. Indeed, the plight of the urban poor has lately worsened, mainly because of the severe drought in 1992 and the recent economic structural adjustment programmes. The situation has also been aggravated by the rapid growth of Harare, which has experienced a 10 per cent annual growth rate in the period 1982–92. In the main, this has been due to the arrival of low-income migrants from rural areas (CSO, 1982, 1992).

The effects of the drought on the country as a whole included the following:

1 Numerous cattle deaths occurred on both commercial land and communal land, with whole herds being lost in many parts of the country (*The Herald*, 12/8/92a, 19/8/92).

2 Food supplies were rapidly depleted (*The Herald*, 13/8/92a, 25/8/92), with consequent hunger, starvation and widespread malnutrition. This required costly emergency feeding programmes (*The Herald*, 13/8/92b, 14/8/92a, 22/8/92, 24/8/92).

3 Water sources dried up, affecting livestock, people, commerce and industry (*The Herald*, 21/8/92a, 31/8/92a, 31/8/92b). Water rationing and fines for excessive consumption of water were implemented (*The Herald*, 15/8/92, 29/8/92a), and hydroelectric power costs increased (*The Herald*, 14/8/92b, 20/8/92).

4 In spite of emergency relief aid from abroad, the national debt increased as a result of government loans to support the agricultural sector, to cover costs of food imports and their distribution, and to fund supplementary feeding programmes (*The Herald*, 12/8/92b, 18/8/92a, 21/8/92b, 29/8/92b).

5 Inflation increased, as production costs rose due to escalating fuel costs and scarcity of goods (*The Herald*, 6/8/92a).

These effects of the 1992 drought aggravated the general economic hardship brought on by a structural adjustment programme, implemented in 1991 at the behest of the World Bank and International Monetary Fund. As elsewhere in Sub-Saharan Africa where such programmes have been adopted, its main features were (a) trade liberalization, including devaluation to encourage exports, removal of controls on imports and on foreign exchange and investment, decontrol of prices, increased interest rates, and a transfer of taxes from production to consumption, (b) streamlining of the public sector, notably by wage freezes, privatizations, and cost recovery measures on public services, and (c) subsequent debt restructuring in order to encourage aid and investment (Seralgeldin, 1989; Government of Zimbabwe, 1991).

Although some economic growth has resulted from the adjustment programme (Seralgeldin, 1989; Jesperson, 1992), increasing economic hardship
has also been evident. For example, the reduction in public sector employment has not been compensated by private employment; local industries
have been undermined by trade liberalization; average and minimum wages
have fallen in real terms; consumer demand has been dampened; public
expenditure on health, education, nutrition and other welfare services has
been cut; and cost recovery measures (i.e. charges) have been introduced
(Seralgeldin, 1989). Specific indications of worsening economic conditions in
the early 1990s include a decline in gross domestic product (EIU, 1993a),
annual inflation in the range 25 to 45 per cent (EIU, 1993b), and lower wages
in real terms (Balleis, 1993). There has also been a reduction in manufacturing
production, leading to lay-offs and closures that have increased unemployment (ZFH, 1993; Rakodi, 1994). *Per capita* expenditure on public health and
education fell in real terms by 20 per cent and 14 per cent respectively
between 1991 and 1992 (Balleis, 1993).

It is not entirely clear whether the drought or the structural adjustment
programme has been the more damaging to the population, but there is no
question that the economic hardship of the urban poor has been greatly
aggravated by declining real wages, growing unemployment, and the increasing costs of services.

Of all the needs of the urban poor, food is the most fundamental. According to Drakakis-Smith and Kivell (1990), low-income families in Harare
typically spend more than 70 per cent of their income on food (cf. with only
10 to 40 per cent for high-income families). In the early 1990s, food prices
have risen dramatically. In August 1992, for example, the price of staple
maize meal increased by 50 per cent in one day, and that of bread by 65 per
cent, as a result of the removal of price subsidies (*The Herald*, 6/8/92b). Such
events threaten the ability of low-income families in particular to obtain
adequate food, and necessitate a search for alternative food sources. Growing
their own food is one possibility, and this has contributed to the steady
increase in cultivation of public land in Harare, in spite of the efforts of the
municipal authorities.

Policy and Planning Responses

The produce of urban cultivation has been a vital support for low-income
families in Harare during the recent severe drought, and has partly coun-
tered the negative effects of the structural adjustment programme. Nevertheless, the municipal authorities have continued to destroy crops grown in
prohibited areas. Indeed, it appears from perusal of the City Council minutes

that the more the population has cultivated public land, the more the authorities have been determined to slash the crops, as if cultivation were a deliberate and wanton act of defiance.

Given the current global concern with environmental issues, the attempted control of cultivation on public land for reasons of environmental protection can be commended. However, the control has increased hardship among the urban poor and provoked conflict between them and the authorities. Given that, as yet, no information exists on the actual environmental impact of such cultivation, the current attempts at control are based more on conjecture and hearsay than substantiated need. In view of the human consequences of the attempted control, its practicality is questioned. In contrast to the official response in Zimbabwe, in parts of Asia and the Pacific, for example, the benefits of urban cultivation, as a means of providing a more varied and nutritious diet, are widely extolled and positively encouraged (Thaman, 1984).

In these circumstances, and, given the difficulty of finding alternative means of meeting food needs in Harare, it has been clear for some time that the current policy is in need of review. One possibility discussed in the mid-1980s was that plots of land should be designated by the authorities for cultivation by city residents on a co-operative basis. It was hoped that such plots would satisfy the needs of residents as well as being more susceptible to official control. The Department of Town Planning and Works drew up maps of land in Harare suitable for cultivation. The land was mainly of low relief, and included some *vleis* (Department of Works, 1984). The Department of Health, Housing and Community Services (DHHCS) assumed responsibility for allocating the plots. Allocations were made on the understanding that plots were returnable to municipal control if they were needed for other purposes, e.g. housing development. Since the mid-1980s, several co-operatives of this kind, each covering a few hectares, have been established, mostly in high-density suburbs like Dzivarasekwa, Kuwadzana, Kambuzuma, Mufakose, Mbare and Mabvuku.

Co-operative cultivation of this kind has been heavily criticized and often considered unsuccessful. In places, individuals have been unable to find suitable neighbours with whom to establish a co-operative, because of the transitory status of many residents and their diverse origins and agricultural experience. Even where established, such co-operatives have often later collapsed due to poor communication and understanding among participants or an inability to raise capital to cover the costs of inputs. Problems have also arisen over paying the charges that the DHHCS sometimes levies. The charges are a particular disincentive for the low-income families who would most benefit from participation. The lack of security of tenure also reduces the investment that the participants are willing to make, and the scheme is often unpopular with non-participants who may have lost access to cultivable land. In one case, an established co-operative had its crops destroyed by official 'slasher squads', and is now suing the DHHCS for compensation (*The Herald*, 5/2/92).

The establishment of co-operative schemes, which are more easily controlled by the authorities, has not resolved the problems of the urban poor. Indeed, they threaten to exacerbate them by being no more environmentally sound than other cultivation and by adding credit problems to the issue. Modifications to the schemes are required, if they are to resolve the persisting conflict between the urban poor and the municipal authorities.

Conclusion

The cultivation of public land in Harare provides a vital, supplementary food source for the urban poor and relieves demand on their limited income. It has been an important self-help strategy during the recent economic hardship associated with the drought and with economic structural adjustment. At the same time, there has been growing conflict between the urban poor and the municipal authorities because of attempts to control the cultivation. The conflict has been aggravated by the multitude of institutions and officials involved, which has resulted in inconsistent and contradictory actions. The real reasons behind attempts to control cultivation seem to have been forgotten by the authorities, who also fail to understand the implications of their actions for the urban poor.

In order to improve the situation for all concerned, further research into the cultivation of public land is required, particularly in regard to the environmental degradation it causes, the extent to which degradation might be lessened by other means than prohibiting cultivation, and the possibility of satisfying the need for cultivation in other ways. The motives for seeking to control cultivation need to be re-evaluated, and the control practices adjusted in the light of the findings.

Acknowledgements

The study was undertaken while the author was a visiting lecturer in the Department of Geography, University of Zimbabwe and funded by the British Council. Thanks are due to Dr D. Tevera, Chairman of the department, for his help and encouragement, and also to the municipal officials, staff in the National Archives, and residents of Harare who assisted during the study. The award of an overseas conference grant from the British Academy is gratefully acknowledged.

References

Balleis, P. (1993) *A Critical Guide to ESAP*. Gweru: Mambo Press.

Bell, M., Faulkner, R., Hotchkiss, P., Lambert, R., Roberts, N. and Windram, A. (1987) *The Use of Dambos in Rural Development, with Reference to Zimbabwe*. Unpublished Final Report of ODA Project R3869, Loughborough University and University of Zimbabwe.

Cormack, J.M. (1972) Efficient utilisation of water through land management. *Rhodesia Agricultural Journal*, **69**, 11–16.

CSO (1982) *1982 Census*. Harare: Central Statistics Office.

CSO (1992) *1992 Census*. Harare: Central Statistics Office.

Department of Works (1984) *Proposed Sites for Allotments*. Plan No. TPR472. Harare: Planning and Development Division, Department of Works.

Drakakis-Smith, D. and Kivell, P. (1990) Urban food distribution and household consumption: a study of Harare. In A.M. Findley, R. Paddison and J.A. Dawson (eds), *Retailing Environments in Developing Countries*. London: Routledge, pp. 156–80.

EIU (1993a) *Zimbabwe Report, September 1993*. York: Economics Information Unit, Centre for Southern African Studies, University of York.

EIU (1993b) *Zimbabwe Report, January 1993*. York: Economics Information Unit, Centre for Southern African Studies, University of York.

Elwell, H.A. (1983) *Notes on Conservation Aspects of Vlei Use*. Borrowdale: Institute of Agricultural Engineering.

Elwell, H.A. and Davey, C.J.N. (1972) Vlei cropping and the soil and water resources. *Rhodesia Agricultural Journal Technical Bulletin*, **15**, 155–68.

Gore, C., Katere, Y. and Moyo, S. (eds) (1992) *The Case for Sustainable Development in Zimbabwe – Conceptual Problems, Conflicts and Contradictions. Report Prepared for the United Nations Conference on Environment and Development*. Harare: ENDA-Zimbabwe and ZERO.

Government of Zimbabwe (1991) *Zimbabwe: A Framework for Economic Reform (1991–95)*. Harare: Government Printer.

Grant, P.M. (1974) What is necessary for wetland maize? *Rhodesia Farmer*, **45**, 43–5.

The Herald (Harare) (1982) Council clamps on illegal cultivation, 4 August.

The Herald (Harare) (1982) City cops to watch for river bank crops, 23 October.

The Herald (Harare) (1982) Council to slash crops on streams, 12 November.

The Herald (Harare) (1984) City growers defy council warning, 8 November.

The Herald (Harare) (1984) Streambank cultivation on the increase in Harare, 16 December.

The Herald (Harare) (1984) Maize growers stage demo after crops destroyed, 29 December.

The Herald (Harare) (1985) 5000 gardens destroyed, 5 January.

The Herald (Harare) (1985) Streambank cultivation tour ends in fiasco, 6 September.

The Herald (Harare) (1985) Streambank slasher squads strike, 7 November.

The Herald (Harare) (1986) Crop slashing will continue says Kanengoni, 18 February.

The Herald (Harare) (1989) Council to slash illegally cultivated crops, 2 November.

The Herald (Harare) (1990) Council to stop illegal cultivators, 7 June.

The Herald (Harare) (1990) Streambank cultivation rife, 28 November.

The Herald (Harare) (1990) Illegal farming worries council, 30 December.

The Herald (Harare) (1991) Illegal cultivation warning, 23 November.

The Herald (Harare) (1991) Council workers slash crops on streambanks, 18 December.

The Herald (Harare) (1991) Uproar over maize slashing, 29 December.

The Herald (Harare) (1992) Slashing warning, 8 January.

The Herald (Harare) (1992) Municipal workers slash maize of Mufakose co-op, 5 February.

The Herald (Harare) (1992) City plans strategy on illegal cultivation, 1 March.

The Herald (Harare) (1992a) Drought raises alcohol prices, 6 August.

The Herald (Harare) (1992b) Shock rise in price of meal and bread, 6 August.

The Herald, (Harare) (1992a) Cattle starving to death in Matabeleland South, 12 August.

The Herald (Harare) (1992b) President assures food worth $1.2bn will last to harvest, 12 August.

The Herald (Harare) (1992a) Region faces critical cereal shortage, 13 August.

The Herald (Harare) (1992b) Seke malnutrition up, 13 August.

The Herald (Harare) (1992a) Five million people register for food aid, 14 August.

The Herald (Harare) (1992b) CFU calls for power rationing, 14 August.

The Herald (Harare) (1992) Harare to impose water fines, 15 August.

The Herald (Harare) (1992a) State to spend $52 million on drought relief food, 18 August.

The Herald (Harare) (1992b) Higher/lower income gap widens in urban areas, 18 August.

The Herald (Harare) (1992) Over 155 000 cattle die in communal areas, 19 August.

The Herald (Harare) (1992) Zesa may soon start rationing of power, 20 August.

The Herald (Harare) (1992a) Marondera authorities panic as water crisis deepens, 21 August.

The Herald (Harare) (1992b) Bank draws $170m to aid farmers, 21 August.

The Herald (Harare) (1992) Drought takes toll on Chihota farm, 22 August.

The Herald (Harare) (1992) Supplementary feeding benefits 1000 children under five in Rushinga, 24 August.

The Herald (Harare) (1992) Mwensi people desperate for food and water, 25 August.

The Herald (Harare) (1992a) Harare needs to save water now, 29 August.

The Herald (Harare) (1992b) 3 Western states give aid worth $95 million, 29 August.

The Herald (Harare) (1992a) Budiriro residents face water crisis, 31 August.

The Herald (Harare) (1992b) 50 000 Rushinga villagers survive on shallow wells, 31 August.

Jesperson, E. (1992) External shocks, adjustment policies and economic and social performance. In G.A. Cornia, R. van der Hooven and T. Mkandawire (eds), *African Recovery in the 1990s: From Stagnation and Adjustment to Human Development*. New York: St Martin's Press, pp. 9–90.

Mazambani, D. (1982) Peri-urban cultivation within greater Harare. *Zimbabwe Science News*, **16** (6), 134–8.

Ministry of Environment and Tourism (1993) *Prospectus for Environmental Assessment Policy in Zimbabwe*. Harare: Government of Zimbabwe.

Moyo, S., Robinson, P., Katere, Y., Stevenson, S. and Gumbo, D. (1991) *Zimbabwe's Environmental Dilemma: Balancing Resource Inequities*. Harare: ZERO Press.

Rakodi, C. (1994) Recession, drought and urban poverty in Zimbabwe: household coping strategies in Gweru. Paper presented at conference of the Institute of British Geographers, 4–7 January 1994, Nottingham.

Rattray, J.M., Cormack, R.M.M. and Staples, R.R. (1953) The vlei areas of Southern Rhodesia and their uses. *Rhodesia Agricultural Journal*, **50**, 456–83.

Scoones, I. and Cousins, B. (1991) *Wetlands in Drylands: The Agroecology of Savannah Systems in Africa. Part 3f: Key Resources for Agriculture and Grazing: The Struggle for Control over Dambo Resources in Zimbabwe*. London: International Institute for Environment and Development.

Seralgeldin, I. (1989) *Poverty, Adjustment and Growth in Africa*. Washington DC: World Bank.

Thaman, R. (1984) Urban agriculture and home gardening in Fiji: a direct road to development and independence. *Transactions and Proceedings of the Fiji Society for the Years 1978 to 1980*, **14**, 1–28.

Theisen, R.J. (1975) Development in rural communities. *Zambezia*, **4**, 93–8.

Thompson, J.G. (1972) What is a vlei? *Rhodesia Agricultural Journal Technical Bulletin*, **15**, 153–4.

Whitlow, R. (1983) Vlei cultivation in Zimbabwe. *Zimbabwe Agricultural Journal*, **80**, 123–35.

Whitlow, R. (1989) *Gullying within Dambos, with Particular Reference to Communal Farming Areas of Zimbabwe*. Unpublished PhD thesis, University of London, London.

Windram, A. (1983) Small sources have a large potential. *World Water*, June, 36–7.

ZFH (1993) *Zimbabwe Economic Review*. Harare: Zimbabwe Financial Holdings.

19

Urban Degradation in Guyana: Causes, Problems and Policy Options

Patrick Williams

It is possible for cities to provide healthy and stimulating environments for their inhabitants without imposing unsustainable demands on natural resources and ecosystems (Hardoy *et al.*, 1992). For this to be achieved, however, priority must be given to healthy living and working conditions for the urban community. Such goals need to encompass water supply, sanitation, solid-waste disposal, roads, housing, and other infrastructure and services that are necessary for a prosperous urban economic base.

In Less Developed countries, urbanization and industrialization are still at an early stage, and, for this reason, the potential for pollution and other forms of environmental degradation is arguably much greater than in countries with more developed economies. However, rapid urbanization and industrialization need not create serious environmental problems, provided there is an adequate understanding of the environmental implications of such processes and the timely creation of an institutional framework to address the problems.

In Guyana, as in other Third World countries, the process of industrialization and urban expansion is rapid. However, there is a lack of financial resources and in many areas institutional structures are in decay, which in turn leads to deterioration of the urban infrastructure and reduction in the quality of the urban environment. Against this background, the present chapter examines issues relating to urban environmental quality in Guyana, particularly with regard to water quality, drainage, solid-waste disposal and housing. Some policy implications are explored in the latter part of the chapter. Attention is focused on the towns of Georgetown, New Amsterdam, Linden, Corriverton and Rose Hall: the main urban centres in the country. One other urban area, Anna Regina, which is relatively small, has only recently been accorded urban status (Figure 19.1).

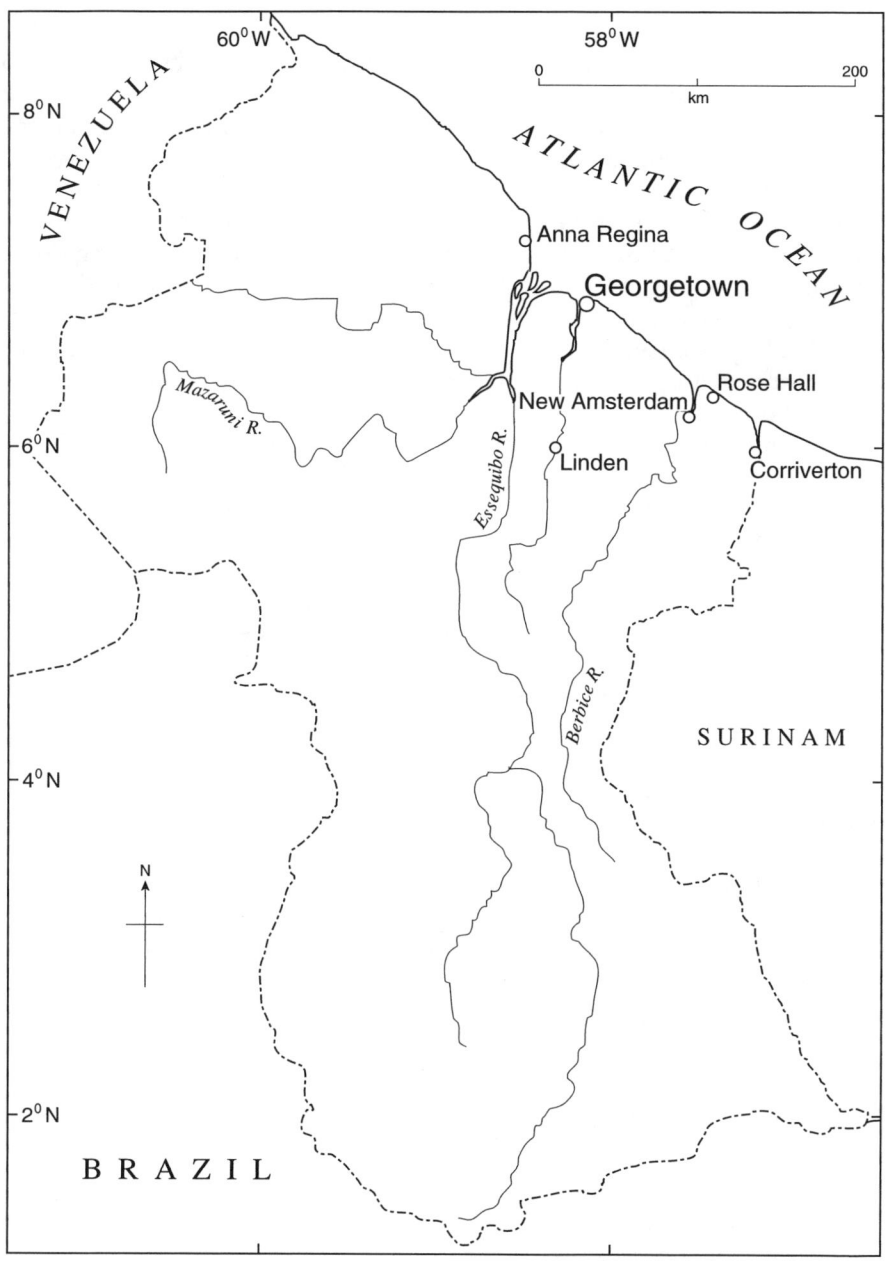

Figure 19.1 Location of urban centres in Guyana.

Institutional Framework

Urban settlement in Guyana dates back to the 1780s when the settlement of Long Champs, later to become the national capital of Georgetown, was laid out by the French. Wards were subsequently added with rectangular grid patterns on the European model. However, minimal attention was given to urban planning until the mid-1940s, when a serious fire destroyed parts of the Wortmanville area of Georgetown. This led, in 1948, to the Town and Country Planning Ordinance, which was enacted to control building development and land use in Guyana. Subsequently, in 1961, a Town and Country Planning Department was created to prepare schemes that had been authorized by the Central Housing and Planning Authority. At this time, significant growth of the capital Georgetown was occurring in both eastern and southern directions.

Since Guyanese independence in 1966, the system of local government in the country has evolved from that established in the United Kingdom. The Municipal and District Councils Act of 1969 sets out the basic powers and functions of such councils. However, these powers and functions were subsumed in the Local Democratic Organs Act 1980, which gave Regional Democratic Councils broad powers of approval and monitoring in respect of local council budgets, expenditure and capital development. Until the late 1970s, the Municipal Councils in Guyana functioned as relatively efficient suppliers of services and collectors of local revenue. However, as a result of subsequent economic decline, only minimal services have been made available in urban areas.

The present system of government in Guyana has three tiers, namely, the national, regional and local. At the local level, municipalities are responsible for roads and land drainage, water supply and sanitation, solid-waste collection, environmental health, the operation of markets and abattoirs, and the collection of rates and taxes. Also there are a number of sectoral agencies with responsibilities for specific services in urban and other areas. These include the Guyana Electricity Corporation, the Central Planning and Housing Authority, and the Ministry of Health.

Causes of Urban Degradation

The urban centres of Guyana are located on the coastal plain and on the interior sand belt. Georgetown, New Amsterdam, Corriverton and Rose Hall (the urban centres of the coastal belt) lie below sea level and are characterized by low relief, saturated soils and a shallow water table. They are prone to various environmental problems relating to: (a) the supply of suitable drinking water; (b) sanitation (in both private, residential and public buildings); (c) solid-waste disposal; (d) land drainage; (e) the quality of public areas; and (f) public safety. In contrast, the town of Linden (located on the interior sand belt) is built on permeable sandy soils and the water table is deep. Some of the general problems of the urban centres of the coastal plain recur, but Linden's main problems are air quality and soil erosion.

Economic, political and administrative problems have contributed to the deterioration of the urban infrastructure in Guyana. For much of the 1970s and 1980s, the public sector expanded into practically all areas of the economy including manufacturing, trading and financial services and also the main export sectors of sugar, bauxite and rice. The bureaucratization of production was an explicit attempt to create a socialist economy, but the approach proved inimical to national development. The economy was undermined by various factors, including weak public sector management, poor programming and budgeting, falling prices for sugar and bauxite, and the international oil crisis of the mid-1970s. Between 1976 and 1988, the gross domestic product of Guyana declined by 32 per cent.

By the late 1980s, the economic situation was so severe that dereliction and environmental degradation were widespread, and the economic and social infrastructure was on the verge of collapse. In 1988, the government formulated an Economic Recovery Programme, which was introduced in the 1989 budget. The programme was a major shift in policy, seeking to reduce state involvement in the productive sector by means of privatization and a return to a free-market economy. The programme sought to provide incentives for private investment in areas like agriculture, mining and forestry, and to rehabilitate the infrastructure and services that had severely deteriorated.

While significant economic improvements have been achieved, with recent growth rates exceeding 5 per cent *per annum*, such have not yet benefited the urban infrastructure. A report from the Urban Rehabilitation Programme (Anon., 1993a, 1993b) confirmed that municipal services and management in Guyana, albeit relatively effective until the late 1970s, deteriorated dramatically thereafter. This reflected the general economic decline, which in turn led to foreign exchange restrictions and the curtailment of imports, notably of new plant and equipment, as well as machinery spares. The problem was compounded by several currency devaluations, which greatly increased prices and reduced real wages. These national economic woes contributed significantly to the problems of urban areas.

At the municipal level, the dominance of the public sector also adversely affected the urban economy. From the late 1980s, many municipalities became increasingly dependent on central government for support, and, with cut-backs in government expenditure, levels of investment in urban infrastructure were severely limited. Likewise, private sector investment contracted during this period. Many households suffered severe economic hardship and found it virtually impossible to pay the requisite rates and taxes which resulted in a further deterioration in services. In April 1993, it was estimated that the actual costs of maintaining and operating the main essential services in the five municipalities were some G$283 million (*c*.£1.4 million at 1994 rates), a sum which was well beyond the resources of the municipalities (Table 19.1).

Table 19.1 Estimated annual maintenance and operating costs of the main essential services in municipal areas in Guyana in 1993.

	Roads	Drains	Street lights	Sanitation	Markets	Sewage and water supply	Total
				(million G$)			
Georgetown	32.97	22.78	8.36	9.38	12.00	55.0	140.49
Linden	6.09	3.64	0	3.30	0.75	10.0	23.78
New Amsterdam	37.32	20.75	1.92	3.97	3.00	7.5	74.46
Corriverton	12.09	7.23	0.16	1.82	2.00	5.0	28.30
Rose Hall	6.09	3.64	0.19	0.93	0.25	5.0	16.10
Total	94.56	58.04	10.63	19.40	18.00	82.5	283.13

Note: G$200 = *c*.£1 at 1994 rates.
Source: Anon. (1993b).

Even now, in an improving economic climate, the municipalities still struggle with central government to increase their levels of subsidy. It is clear to many observers that political considerations now influence government's disbursements to the municipalities. Those councils that are dominated by opposition members seem to be deliberately starved of cash, while the converse applies to councils that support the ruling government.

Other political factors have also affected the status of urban areas. In general, there has been a lack of interest in local politics, which has been evident in the reluctance of individuals to stand for municipal office in towns such as New Amsterdam and Corriverton (Anon., 1993a, 1993b). It was generally felt that the councils were not democratically elected, replicating a situation at the national level. The councils rarely supported or participated in activities aimed at community development. In addition, councils suffered from a lack of accountability, primarily because councillors were elected on a party slate. Many councillors were ignorant of municipal laws and regulations, a situation which further impaired the effectiveness of the councils.

A number of administrative problems are also apparent in urban areas. They include low staff salaries, shortages of qualified personnel, lack of staff training schemes, lack of modern office equipment, and poor quality office accommodation. Local government salaries, for example, range from G$1050 per month (*c*. £5 at 1994 rates) for typists and clerks at Linden to a maximum of G$18 675 per month (*c*. £95) for the Town Clerk in Georgetown. Low salaries and the lack of qualified applicants make it difficult to recruit suitable staff to councils, and explain high staff vacancy rates, particularly in Linden and Corriverton (Table 19.2). These factors, plus the low morale among existing staff, affect the ability of municipalities to undertake their statutory functions.

Table 19.2 Municipal staffing and salary levels in 1991 in Guyana.

Council	Population (1991)	Authorized staff size	No. of vacancies	Salary scale (G$ per month)
Georgetown	158 315	1388	170	4000–18 675
Linden	26 820	326	172	1050–10 000
New Amsterdam	18 622	193	52	3136– 9 940
Corriverton	12 900	79	27	3360–10 080
Rose Hall	5200	40	6	3303–12 000

Note: G$200 = c.£1 at 1994 rates.
Source: Anon. (1993b).

The ineffectiveness of municipal government is illustrated by its inability to generate council revenue. With the exception of Georgetown, there have been no rating revaluations in urban areas since the late 1970s. This is in violation of the Valuation and Rating Purposes Act 1977, which makes provision for revaluations every five years. As a result, rateable values in many urban areas are currently too low, reducing the potential income of councils. The problem is compounded by the ineffectiveness of rate collection, which results in all towns having significant sums of money in arrears, in spite of their desperate financial situation. The problem is particularly acute in Georgetown, where the town council is on occasions obliged to borrow money from commercial banks at high interest rates in order to cover its liabilities.

Urban Environmental Degradation

In Third World countries a significant proportion of the population commonly lives and works in very poor conditions. In many Third World cities, between 33 per cent and 66 per cent of the population occupy inadequate housing units (Hardoy *et al.*, 1992). There are inadequate piped water supplies and inadequate provisions for the removal of sewage and other wastes. Proper land drainage is generally lacking, as are all-weather sealed roads. Sewage systems are often non-existent or unsatisfactory. In Accra, Ghana, only 30 per cent of the population is reportedly served by waterborne sewerage; comparable figures for Jakarta, Indonesia and Bangkok, Thailand are 6 per cent and 2 per cent respectively. Likewise, in Kampala, Uganda, 81 per cent of the population depend on pit latrines; a similar high dependence (71–98 per cent) is reported for major Nigerian cities like Benin, Calabar, Kaduna, Kano, Ibadan, Sokoto and Zaria (Hardoy *et al.*, 1992). Pit latrines have long been considered a health hazard because they harbour flies. There is the additional disadvantage of their unpleasant odour. As such, they reduce the quality of the urban environment. Guyanese towns suffer comparable problems.

Water and Sanitation

In Guyana, the towns of the coastal zone generally obtain their water supplies from shallow wells, while Linden's water comes from the upper course of the Demerara river. In Linden, Georgetown and New Amsterdam, water is treated and is initially of acceptable quality for domestic use. No water treatment is provided at Rose Hall or Corriverton. The major problem of water supply in all these towns is the vast number of leak points in cracked pipes. These commonly lead to water contamination and related health problems, especially in the rainy season, when urban areas are often flooded.

Table 19.3 Urban water supply in 1991 in Guyana.

Council	Water supply connections		Source	Quality of distributed water
	House	Other		
Georgetown	25 000	350	Boreholes (12), reservoir (1)	Poor
Linden	n/a	n/a	Borehole (1), river intakes (4)	Good
New Amsterdam	2980	432	Boreholes (2)	Poor
Corriverton	1300	227	Boreholes (3)	Fair
Rose Hall	780	49	Boreholes (3)	n/a

Source: Anon. (1993b).

In Table 19.3, data are presented on the water supply system in the urban areas. They indicate that only Linden has water of good quality after account is taken of the distribution network. According to the report on the Urban Rehabilitation Programme (Anon., 1993b), the deterioration in the network is attributable to underfunding of the water supply system over recent years.

Georgetown is the only town in Guyana which is partially served by water-borne sanitation. The effluent from the system is untreated and is currently discharged into the Atlantic Ocean. At present, the sewerage system in Georgetown is in very poor condition and it is not uncommon to observe back flow of raw sewage seeping through manholes. Elsewhere in George-town and in other towns, the population is served by on-site sanitation, either in the form of septic tanks and filter boxes or pit latrines. These systems are generally impaired by the prevailing clayey subsoils, which inhibit effluent percolation and encourage discharge into surface drainage. Of the total urban population of some 222 000 in Guyana in 1991, only 22 per cent (all in

Georgetown) were served by water-borne sewerage. The majority (52 per cent) had septic tanks, while 21 per cent were served by pit latrines (Table 19.4).

Table 19.4 Urban sanitation in 1991 in Guyana.

Council	Population	Sanitation (by population)			Other facilities
		Sewerage system	*Septic tank*	*Pit latrine*	*(public toilet)*
Georgetown	158 315	49 000	103 000	6000	8
Linden	26 820	0	4250	12 750	2
New Amsterdam	18 622	0	3730	14 932	3
Corriverton	12 900	0	4515	7740	0
Rose Hall	5 200	0	500	4500	0

Source: Anon. (1993b).

In Guyana septic tanks and pit latrines are a particular health risk, because of the low elevation of the coastal towns and their susceptibility to frequent flooding. It is not uncommon to encounter raw sewage in housing areas during floods. There is also the hazard of discharge from septic tanks flowing into open drains within the urban areas.

Table 19.5 Urban solid-waste management in 1992 in Guyana.

	Collection frequency			Collection equipment	
	Central business district	*Residential area*	*Market area*	*Truck*	*Tractor and trailer*
Georgetown	weekly	2× month	daily	18	1
Linden	weekly	none	weekly	4	1
New Amsterdam	3× month	3× month	daily	1	3
Corriverton	2× month	2× month	daily	0	1
Rose Hall	weekly	weekly	daily	0	1

Source: Anon. (1993b).

Solid Waste

Solid-waste management in urban areas has suffered greatly over the years from underfunding and neglect, and is currently both inadequate and inefficient (Table 19.5). In 1992, for example, the number of vehicles available for garbage collection in Georgetown was reduced to three from a nominal fleet of twenty-three. Consequently, the municipality had to contract out solid-waste collection services. In view of the additional cost involved, savings have been sought elsewhere, including the dumping of garbage in gap sites within residential areas rather than outside the town.

This practice has serious health implications for urban residents. Firstly, such dumping sites become breeding grounds for rats and flies, with attendant disease vectors. Secondly, garbage dumped in this way often blocks drainage systems, which consequently overflow during the rainy season, creating further health hazards. Thirdly, no effort is made to separate domestic and industrial garbage, so that harmful chemical contaminants from the latter are present in some dump sites. Such areas are often used as playgrounds after they have been filled and covered. The problems are most acute in Georgetown, but occur in the other urban centres.

Housing Quality

A very high proportion of dwellings in the urban centres are run-down and overcrowded. The problem is most acute in Georgetown and Corriverton. Approximately 52 per cent of the dwellings in Georgetown and 64 per cent in Corriverton are over 30 years old, and most have received little or no maintenance in recent years. The dilapidated buildings are unsightly, and also present serious fire hazards. In poor sections of central Georgetown, where the percentage of old buildings attains 63 per cent, extreme overcrowding is also characteristic (Table 19.6).

Table 19.6 Urban housing in Guyana in 1980.

	Total units	Privately owned	Detached units	Built before 1961
		(%)	(%)	(%)
Georgetown				
old town	12 964	26.4	33.9	63.1
suburbs	22 328	34.7	44.6	46.3
total	35 292	31.7	40.7	52.5
Linden	6389	40.8	37.5	18.7
New Amsterdam	4142	39.8	64.6	45.5
Corriverton	2688	56.3	70.4	63.6
Rose Hall	1410	68.8	84.6	39.0

Source: Anon. (1993b).

In recent years, squatting has contributed to the degradation of urban environments. In the case of Georgetown, squatter settlements have commonly been established alongside canals that supply clean water to the town. It is now apparent that domestic waste is increasingly being dumped into these canals.

Problems in Linden

Linden, with a population of 26 820, is the second largest town in Guyana and the site of several bauxite plants. It lies in the interior sand belt, and has special environmental problems. Firstly, soil erosion is a problem on account of the local 'sand hill and valley' relief, particularly in the Wismar area. Here the vegetation cover has been removed from hill slopes to facilitate housing developments and to provide fuelwood. As a result, large quantities of sand are washed into adjacent valley bottoms during the rainy season, in places engulfing the lower levels of buildings. Secondly, air pollution from bauxite processing plants in the vicinity of the town is a concern for residents. Pollutants, mainly in the form of dust, are carried from the chimneys of the bauxite plants by the prevailing winds to the poor sections of the town, mainly Wismar. As a result, buildings always appear in a shoddy condition and the surrounding vegetation suffers. Such pollution is considered to have adverse effects on residents with bronchial ailments, although no statistical evidence is currently available to support this.

The problems at Linden are compounded by the fact that bauxite, which is the mainstay of the urban economy, has lately been in low demand. In addition, there are local management, production and productivity problems. The industry has cut back on its operations and also its workforce. The dramatic rise in unemployment in the town has directly affected the whole urban economy and consequently rate collection. This has in turn limited the ability of the municipal authorities to maintain the urban infrastructure and services.

Conclusion

Severe environmental degradation exists in urban centres in Guyana. It is primarily associated with a build-up of garbage, blocked drainage systems, dilapidated and overcrowded housing, and, in the case of Linden, with soil erosion and air pollution. The degradation reflects underlying socio-economic and administrative problems that have confronted both municipal administrators and residents. In turn, these problems are the result of years of neglect and maladministration in the municipalities. In order to resolve such problems, several policy initiatives are required:

1   Environmental education. There is an urgent need to educate urban residents about the issues of sewage and garbage disposal and the associated health implications.
2   Training in urban management. There is a need to recruit and train skilled personnel for urban management. At present, municipalities lack the highly qualified staff to undertake the necessary planning, programming

and budgeting. Underlying this is the need to improve staff wages and salaries and also the working environment in order to attract suitable personnel.

3   Financial resources. Lack of finances is a major cause of the problems outlined. The inability to collect garbage, clear drains, and repair sewers reflects inadequate financial resources. One solution to this would be to improve property valuation services, but it has to be recognized that the residents' ability to pay adequate rates is compromised by prevailing low income levels.

4   Compensation payments. It is essential that where private bodies cause environmental degradation, they should pay for their actions. This applies particularly in the case of mining companies at Linden, where sand erosion and air pollution are evident.

5   Integrated planning. All the municipalities need to prepare their programmes within the national framework, so that activities with broader implications can be accommodated. At present, municipalities make *ad hoc* decisions that are commonly based on outdated maps and land use plans. Development-control legislation is likewise outdated.

References

Anon. (1993a) *Urban Rehabilitation Programme. Diagnostic Report*. Georgetown: Inter-American Development Bank.

Anon. (1993b) *Urban Rehabilitation Programme, Stage 1 Report. Vol. II. Infrastructure and Investment Programme, Government of Guyana*. Georgetown: Inter-American Development Bank.

Hardoy, J.E., Mitlin, D. and Satterthwaite, D. (1992) *Environmental Problems in Third World Cities*. London: Earthscan Publications.

20

Environmental Impacts of Urban Development and the Urban Informal Sector in the Caribbean

Sally Lloyd Evans and Robert B. Potter

Throughout centuries of change and development in the small island states of the Caribbean, human interaction with the physical environment has remained highly visible. Indeed, the Caribbean region, which is home to 35 million people, is one of the most highly transformed regions in the world (Potter, 1989, 1993, 1995). Rapid urbanization and recent economic change have substantially altered both the natural and social environments. As Demas (1965) and Worrell (1987) have observed, the contemporary Caribbean is characterized by small territorial size, undiversified economies and economic dependency.

The region has been subject to centuries of imposed external power. This domination has extended through slavery, to the importation of western models of development involving industrialization and modernization (Lewis, 1955), and, lately, to economic restructuring agreements with the International Monetary Fund (Thomas, 1988). Unfortunately, political independence has not ended the external pressures on Caribbean environments. Of late, attention has focused on environmental degradation as a function of colonial policy and of recent activities such as tourism, urbanization and industrial development (Thomas, 1988; Potter, 1989; Deere *et al.*, 1990; McAffe, 1991; Richardson, 1992). Thomas (1988) notes that the term *crisis* is the most widely used to describe prevailing social, economic and environmental conditions in the Caribbean. The crisis can be attributed to growing poverty, economic collapse, the disintegration of domestic food production, inappropriate industrial policies and rapid urbanization.

This chapter explores the nature of environmental degradation in the region, with specific reference to recent industrial, commercial and urban change. Emphasis is placed on the growth of the informal sector of the economy which is presenting the region with a new set of environmental concerns and which is highlighted in a case study from Trinidad.

History of Economic Change and Environmental Degradation

Environmental degradation, as a direct result of economic change, is not a recent event in the Caribbean. Columbus's voyage of 'discovery' represented a spatial expansion of political and economic control (Wolf, 1982) that greatly influenced the natural environment. The Caribbean has been part of the international system of trade for four centuries, a position that is exemplified in the region's integration within the international division of labour. The effects of slavery, emigration and foreign industrial investment are widespread. The Caribbean's external focus results from its historical introduction to trade from the 1400s, the environmental effects of which have been far-reaching.

The islands were heavily forested and the vegetation diverse when Columbus first arrived in the region (Richardson, 1992). The indigenous inhabitants, the Carib and Arawak Indians, disappeared soon after the Spanish conquest due to enslavement and disease (Watts, 1987). Their disappearance marked the first stage of environmental devastation in the region.

To the Spanish, the Greater and Lesser Antilles provided land for imported agricultural techniques and breeding grounds for livestock. As Spanish agricultural staples did not prosper in the Caribbean, Columbus brought in sugar cane, the crop which was to become synonymous with the region. Rapid deforestation of the land for planting cane was undertaken by imported African slaves. The clearance of vast amounts of forest also produced firewood and construction timber, a move which added to the massive environmental change that was transforming the region. Abandoned villages, widespread sugar cane cultivation and savannas used for cattle grazing all led to rapid soil erosion and leaching. Large areas were deforested and, by the mid-1600s, extensive clearance had occurred throughout the Lesser Antilles.

The subsequent conquest of many of the islands by the British, Dutch and French only enhanced the environmental mismanagement, particularly in relation to the plantation economy. Colonization was a gradual process, which started with the cultivation of crops like tobacco, cotton and cocoa, and later sugar cane. Many islands were turned into mono-crop economies, as the plantation competed heavily with local food crops and forest. The islands became absorbed in the expanding European commodity exchange over which they had no control. From the 1600s, the growing European demand for cane sugar gave little regard for the widespread social and environmental devastation that resulted from the plantation system.

The plantation combined factory and field at an early date, and effectively represented 'industry' prior to the rise of industrial production in Europe (Mintz, 1985). The plantation was based on the production of sugar cane through the use of imported slave labour under European management. Plantation agriculture in the region has been well documented by Beckford (1972) and Sheridan (1973), but the ecological impact of the colonial sugar-cane plantation has received less attention than its social and demographic impacts (Richardson, 1992). Yet it is clear that the cultivation of sugar cane, combined with widespread forest clearance, represented an ecological discontinuity with the past. The most spectacular, documented case of environmental devastation in Caribbean history was the burning of the forest and scrub of the entire island of St Croix by the French 'to make the islands more healthy' for the European settlers (Dirks, 1978, p. 16).

The immense influence of the plantation is displayed throughout the Eastern Caribbean in terms of settlement and land use. Rojas (1989) referred to the Caribbean settlement pattern as *plantopolis*, a pattern which is heavily influenced by the plantations. Likewise, land use transformation followed the demand of the plantations. Their requirements stretched beyond the need for cleared land, as large amounts of timber were required for construction and fuel (Sheridan, 1973). The islands were not just subject to physical degradation, but were also biologically transformed. New plants and animals were introduced, many of which added to the destruction of indigenous vegetation. As more land was cleared, European settlers favoured a move to the coast, as in Guyana where sea walls were constructed on mangrove swamps and mud flats. The situation changed little after the emancipation of slaves in the 1830s, when new cane factories and railroads were introduced to aid economic production. In this way, the environment was constantly degraded by uncontrolled economic practices.

The colonial legacy is one of sugar cane domination, supplemented by the export of other staples. During slavery, the local production of subsistence crops had been limited, a problem which has persisted to the present. Many islands still have to import a large proportion of their fruit, vegetables and other staples. The significance of the physical degradation of the environment cannot be overemphasized. Food deficits remain a problem throughout the region, while the inability of households to expand incomes through subsistence cropping, due to infertile soils, is a major difficulty in times of hardship. The colonial powers completely dominated the Caribbean environment in a way that was not seen elsewhere. The land now inherited by the people of the Caribbean has been used to gain profit for centuries in a similar way to the utilization of their labour by foreign powers (Besson, 1987; Potter, 1992). Caribbean degradation is as much a legacy of the past as it is a result of the twentieth century.

Today, external economic control in the form of foreign industrial invest-ment, tourism, and structural adjustment is bringing new environmental problems. Agricultural and mineral exports still depend on external markets, and tourism is crucial to the economic survival of many countries. But the economic benefits from the Caribbean environment still mainly accrue out-side the region (Lowenthal, 1987), whilst the disbenefits still affect the islands themselves. The implementation of structural adjustment packages by the International Monetary Fund restricts public spending across the region, and is causing a marked increase in unemployment and poverty and further degradation of the natural and social environments.

Current Environmental Pressures in the Caribbean

It has been argued by McAffe (1991) that the processes which keep wealth flowing from the South to the North violate the natural environment and undermine the ability to earn a livelihood. In the Caribbean, the growing separation between those who control production in industry and agri-culture, and those who provide the labour precludes the implementation of sustainable development. Similarly, escalating debt often prevents Carib-bean nations from giving the environment due consideration. The daily struggle for survival often takes precedence over more environmentally sensitive, and costlier, practices in employment, housing, industry and agriculture. As in the past, environmental degradation caused by inap-propriate economic and urban activity is clearly visible.

The importance of export agriculture over domestic production is result-ing in the use of pesticides and fertilizers. The inability of many countries to feed their populations is exacerbated by a declining ratio of arable land per head of population, which is a direct correlate of urbanization. Inappropriate use is made of forest and water resources for energy consumption, tourism and industry. The rapid growth of tourism has also led to a construction boom that is depleting sand and gravels. Unchecked urban construction and the clearing of swamps and forest for the tourist industry damage the region's natural beauty and its capacity to produce food. Deforestation, with its multiple ecological consequences in places like Haiti and the Dominican Republic, is now spreading to other islands. The result is seen in the tonnes of topsoil that are washed into the sea every year or cause accidents in haphaz-ard housing developments. Likewise, the dumping of waste, periodic in-dustrial accidents, and the abuse of pesticides threaten the long-term in-tegrity of the natural environment.

In the last century, the Caribbean suffered the adverse effects of agricul-tural degradation due to the plantation economy and slavery. The twentieth century has seen pressure for many islands, including Barbados, Jamaica and Trinidad, to industrialize, a transition that has brought with it problems of waste disposal and pollution. The Caribbean Basin Initiative attempts to solve the region's economic difficulties by promoting manufacturing exports. At present, Caribbean governments encourage foreign investment in manufacturing plants and data-processing companies through such incentives as low wages and the curtailment of environmental policies (Safa, 1990; Pearson, 1993).

Whilst environmental degradation is increasing in the 1990s, Caribbean governments have seen their ability to cope with its consequences eroded. The region is much poorer than it was a decade ago due to its adverse balance of payments, falling real wages, and debt and austerity packages that have increased unemployment and poverty. The pressures have de-stabilized the region, whilst economies are still dependent on the international arena. In their desperation to obtain hard currency to service interest repayments, governments invite investors to fell more forests and replace more food crops with export crops (McAffe, 1991). Furthermore, the insular and small-scale nature of these states makes them ill-equipped to cope with the natural disasters that plague the region. The world is fully aware of the devastating effects of Hurricanes Hugo and Gilbert which brought entire countries to a state of emergency.

Socially, increased poverty and declining provision of services have provoked urban blight and increased crime. Large parts of most Caribbean towns are run-down and neglected, with poor access to water and electricity. Directly linked to the urban blight is uncontrolled urbanization, particularly the growth of the informal sector (Potter, 1993, 1995).

The Informal Sector and the Caribbean Environment

For over three decades, the problem of large-scale structural unemployment has been a central concern in the Caribbean, as elsewhere in the developing world. According to Farrell (1978), Caribbean unemployment reflects a shortage of capital, the malfunction of the labour market, excessive population growth, the importation of inappropriate technologies, and the effects of colonization. Unemployment and underemployment in the region have been growing since the 1980s, with official unemployment exceeding 20 per cent and unofficial estimates reaching 50 per cent (Deere *et al.*, 1990).

One outcome of the inability of many Third World countries to provide sufficient housing and employment for their growing labour force is the growth of the informal sector. The sector plays a major role in providing

employment within the debt-ridden economies of many Third World countries (de Soto, 1989; Portes *et al.*, 1989; Tokman, 1989), including the Caribbean (Thomas, 1988; Rampersad, 1991). In this context, the informal sector
refers to unaccountable and unregistered activities.

In the Caribbean, the informal sector covers a range of retailing, production and service activities (Lloyd Evans and Potter, 1992; Lloyd Evans, 1994).
Its merits, in contrast to formal employment, include independence, flexibility, and the evasion of taxes and bureaucracy. In the Third World context,
de Soto (1989) argues that it is a haven for motivated self-employment and
small business development, as well as jobs for the urban poor.

Street Trading and Informal Markets: Urban Pollution and Congestion

Street hawking is a popular occupation in the region, especially in urban
areas, and involves the retailing of a wide range of goods (Harrison, 1991).
The traditional female 'higgler' or 'huckster' selling fruit and vegetables or
imported goods is a familiar sight across the Caribbean (Katzin, 1960;
Durant-Gonzalez, 1985; Le Franc, 1989). The importance of higglering cannot
be overestimated, as it plays an essential role in food distribution. In Jamaica
in 1975, a government minister concluded that 80 per cent of fruit and
vegetables consumed were sold by higglers (Senior, 1991). In Trinidad, East
Indian traders have modernized the traditional role by selling fruit from
stalls on the country's main highways. Street traders also sell petty commodities from jewellery and crafts to imported cassettes and electrical items.
Inter-island trading of commodities is also popular, often being undertaken
by women in the eastern Caribbean (Phillips, 1985).

Despite the contribution of vending or higglering to low-income communities, negative factors also exist. The trading of goods has long been
regarded by governments as unaesthetic, and has only been tolerated because it provides incomes. According to Caribbean authorities, unauthorized
vending is not strictly legal as it escapes detection, legislation and tax. In
Trinidad, the police have the power to arrest those 'vending on the footpath',
along with vendors displaying goods on railings or buildings, or littering the
streets.

The justification for street clearing reflects concern over litter, public
hygiene, and street congestion. When the working day is finished, vendors
often discard perished fruit, paper and other rubbish on the street. Over time,
large areas of refuse build up in residential areas. These often act as playgrounds, and a source of income for children who collect refuse or sell beer
bottles back to factories. Informal markets are a similar problem, but on a
larger scale. They are common, and, because they are unlicensed, are not

entitled to refuse disposal. Piles of waste attract vermin and add to the urban degradation. The problem of unauthorized waste disposal often extends to river dumping, and encourages water-related disease in low-income residential areas.

A further problem is street hygiene. Inter-island traders usually spend two or three days on one island, sleeping and cooking in market areas with no access to clean water or sanitary facilities. Street food sellers also have no safe access to services or clean water, which raises issues of public safety. Vending on the sidewalks of downtown Port of Spain or Bridgetown adds to the congestion of central urban streets, while crude stalls can be dangerous to the public. The selling of 'mango-chow' or nuts at traffic lights has caused serious traffic accidents involving vendors, many of them children.

Yet informal trading provides a livelihood for the urban poor and is increasing in many islands due to escalating unemployment and economic recession. The environmental impacts of informality are persuading governments to restrict its growth, but such action will have social and economic repercussions for those involved. Rather than pursuing street clearing, Caribbean governments should assist vendors to dispose of their waste, and install services in markets and other public places. Making informal vending more environmentally sensitive only requires modest planning and investment. Unfortunately, the environmental hazards of informal production are a more serious obstacle.

Petty-commodity Production: Toxic Waste and Resource Decline

A worrying environmental problem is the expansion of informal production across the Caribbean, from small-scale industrial and manufacturing units to unregistered garages. The problem is acute in Trinidad and Jamaica (Kirton and Witter, 1993; Lloyd Evans, 1994). Haphazard development, usually without regard to safety and planning regulations, has led to unsafe urban workshops and production units. In an attempt to emulate formal manufacturing, many informal businesses are producing low-cost goods for local communities or for export. The manufacturing of spare parts for vehicles, industrial chemicals and glues, clothing and electrical items are popular businesses. Many producers are self-employed, but a tendency exists for larger companies to employ groups of informal workers to produce cheap goods for export.

In the 1990s, the use of sub-contracted, informal firms by large companies, often multinationals, means that an increasing range of goods is made in sweatshops, using old machinery and illegal practices (Safa, 1986; McAffe, 1991). Sub-contracting is often associated with the clothing and electronics industries, where low-paid female workers are located in small workshops using sub-standard materials, like toxic glues or lead-based chemicals. The workshops often pursue cost-cutting and fail to conform with local legislation. Industrial effluents containing non-biodegradable petro-chemicals are frequently released into domestic water supplies, rivers or the sea. Waste products, including toxic metals and non-biodegradable substances, are dumped on waste ground or discharged into rivers that supply water to local residents.

Non-conforming and poorly constructed houses and workshops are all too frequent in the low-income settlements of Kingstown, Castries, St George's, Kingston and Port of Spain. The irony is that population pressures are creating more low-income houses in exposed locations just when climate change may result in fewer, but more intense, storm events in the region (Potter, 1992). This is evident in the hilly areas that make up the suburbs of Kingstown. On the east coast of St Vincent, hundreds of houses on the beach are also exposed to the full fetch of the Atlantic, the residents having been forced to move there by plantation owners (Potter, 1994). Backyard garages leave oil spills to wash into the ground, while 're-tread' operators abandon tyres to rot in the sun. In the absence of regulation, other businesses discharge waste into the atmosphere. The natural environment is also used as a resource by informal firms; forest areas provide wood for furniture-makers and other craftsmen, and also supply fuelwood.

Despite these problems, the informal sector could become more environmentally responsible, as it is an indigenous industry that uses local resources and skills. Many informal activities, such as bottle, glass and paper collecting, are already based on the principle of recycling. Across the Caribbean, old beer and soda bottles are collected and returned to the factory. In Trinidad, old rum bottles are collected by small informal channa (spiced chick-peas) and confectionery producers, and used as containers for the sale of their products. In addition, the informal sector often uses more appropriate technology, local resources and operates on a smaller scale than large foreign businesses.

However, informal producers need assistance to dispose of waste and to operate safer businesses, without their informal status being compromised. Public education can go a long way in assisting with the preservation of the local environment. The informal sector has created jobs and should be encouraged, even though environmental impacts will increase as the sector expands. This is evident in Trinidad, where uncontrolled urbanization and a growing informal sector are causing visible degradation of the environment.

The Informal Sector and Environmental Degradation in Trinidad

The Republic of Trinidad and Tobago, a dual island state which maintains a varied pattern of social and economic activity, has recently witnessed rapid growth in informal activity. Trinidad benefited from oil revenues in the 1960s and 1970s, but was hit by the oil crash in 1983, which threw its economy into recession. The effects of international debt restructuring, a stagnant economy, escalating unemployment, rising urbanization and increasing poverty currently place Trinidad in a vulnerable position. Trinidad is environmentally diverse, ranging from the tropical forest of the northern coast to the mangrove swamps of Caroni and the south. The population is approximately 1.2 million, with 60 per cent living in urban areas and half a million living in Greater Port of Spain (Figure 20.1).

Trinidad is the only Caribbean country with an important indigenous petroleum industry, which has stimulated economic development and the establishment of oil refineries and energy-based industries (Scotland, 1983). Independence was gained in 1962, and the country became a republic in 1974 when it proceeded to direct investment towards heavy industry. The Point Lisas Industrial Complex in southern Trinidad was created, featuring refineries, a chemical fertilizer plant, iron and steel works and smaller manufacturing industries (Figure 20.1). Between 1974 and 1983, the government reportedly received US$17 billion in oil revenues, which were spent on industry, infrastructure and education. Large-scale urban construction paralleled the industrial boom.

The prosperity was short-lived, however, due to the oil crash in 1983. In 1988, Trinidad came under the auspices of the International Monetary Fund and found its economy controlled by a severe austerity package. The official unemployment rate has been around 20 per cent since 1987, while the unofficial rate is probably double that figure (Central Statistical Office, 1990). In consequence, informal activity and unplanned settlement have expanded, creating new environmental impacts. The informal sector is one of the least documented in the economy, but is becoming one of the most important (Lloyd Evans and Potter, 1992).

Current Environmental Pressures: Degradation and Urban Land Use

Throughout the 1970s, public concern focused on the lack of environmental policy in the oil industry, particularly in the light of frequent oil spillages off the southern coast and the leakage of poisonous effluents into the rivers and coastal zone. The lack of strict environmental legislation in the formal economy is recognized by the government, but is not the only environmental problem the country faces; earlier modernization led to rapid urbanization and industrialization in northern Trinidad, with the informal sector playing a major role. This is exemplified along the burgeoning west–east corridor in the county of St George, which extends from Port of Spain to Arima (Figure 20.1).

The 1982 National Physical Development Plan for Trinidad and Tobago provided a comprehensive overview of national development to the year 2000 (Figure 20.1). According to the plan

> environmental planning and conservation essentially involve the dual and complementary activities of the allocation of land or physical resources to the most appropriate uses and the prevention or reduction of negative side-effects

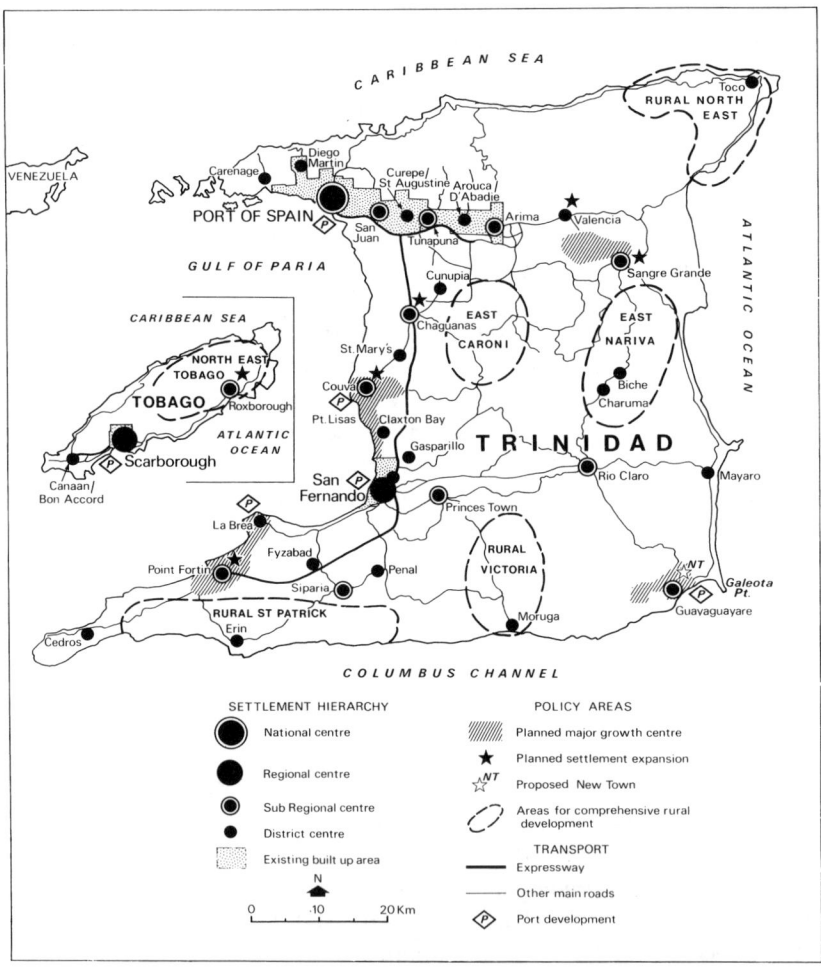

Figure 20.1 Principal features of Trinidad and Tobago, including physical planning proposals for the period 1980–2000.

from positive development activities, which may result in the deterioration of the natural resource endowment and ecological imbalance. (Government of Trinidad and Tobago, 1982, p. 74)

The government admits that there is room for the improvement of regulations, standards, administration and enforcement of environmental policy. The plan also highlighted major environmental problems in Trinidad and

Table 20.1 Environmental problems associated with informal development in Trinidad.

Environmental problem	Causal factors
Damage to fragile ecosystems and watersheds	Lack of legislation/enforcement
Pollution on land and sea	Inadequacy/absence of control
Degradation of landscape	Informal land use
	Informal business development
Denudation of hillsides	Haphazard building
	Small-scale agriculture
Disposal of solid waste and effluents/ health hazards	Industrial, commercial and domestic sources
Land use conflicts	Urban development

Source: Goverment of Trinidad and Tobago (1982).

sought to isolate the causal factors (Table 20.1). Many problems are attributed to urbanization and uncontrolled informal activity, which are still a concern in the 1990s. More recently, the government stated that informal development was a major obstacle to urban environmental sustainability (Government of Trinidad and Tobago, 1991).

Informality and Unplanned Urban Development in Trinidad

Even with improved government control, the majority of informal activities avoid detection and legislation. They include farming, market gardening, building contracting, craft-production, shoemaking, clothing and electrical production, oil-related industry, housing and transport. Small-scale distribution is also popular and encompasses petty traders, street hawkers, sellers of food and drink, bar attendants, agents and dealers. Other services include laundry and vehicle repair (Lloyd Evans, 1994). Although supplying employment, many of these activities contribute to urban environmental degradation in Trinidad.

Uncontrolled development in Port of Spain is a major facet of informal retailing activity. Many residents feel that street trading in Independence Square and informal developments such as the People's Mall on Frederick Street are unaesthetic. Informal markets and street trading of fruit and vegetables also cause pollution. Outside Port of Spain, trading is prominent on the Eastern Main Road, Arima, and in San Fernando. In the numerous street markets on the Eastern Main Road, sanitary facilities or organized rubbish clearance are lacking. Waste products are often left on the street or burnt in large bonfires.

The trading of fresh produce along main highways is popular, particularly amongst East Indian families. Although highway traders provide a valuable service, they are a hazard to motorists and, as well as escaping planning regulations, they forfeit organized rubbish disposal. As a consequence, the highways are frequently littered with unwanted goods. Complaints from residents in the high-income suburb of Valsayn about rotting food and rubbish along the Churchill–Roosevelt highway have recently led to violent street clearances of traders by the authorities. Although statistics are lacking, highway accidents involving vehicles and traders are increasing.

The growth of unplanned settlements on the southern hill slopes of the Northern Range also creates concern. A primary cause is the expansion of Port of Spain into the county of St George. This urban corridor is one of the fastest growing areas in terms of population, housing and informal employment, notably home-based manufacturing and retail and subsistence agriculture. Migrants are moving to unplanned settlements in the corridor due to the improved employment prospects and proximity to Port of Spain (Conway, 1989). As a result of local land shortage, many families are building houses, small farms and workplaces on the surrounding hills. This causes the destruction of forest, flash-flooding, soil erosion and minor landslides, pollution of the San Juan river, and poor environmental conditions in settlements themselves. The remaining forest areas are used as sources of wood for informal craft and furniture-making and for poaching of rare species. Good agricultural land at the base of the forest is also being lost.

Along the Eastern Main Road from San Juan to Arima, the landscape is dotted with automobile junk yards, tyre re-tread centres, scrapyards and rubbish dumps. Informal small-scale production units exist in houses or back-yards in St Joseph and Tunapuna. The services range from welding and car repair to small-scale furniture and appliance repairs. The towns of San Fernando in the south, Arima in the east and Chaguanas in the central region are also witnessing an explosion of informal activity (Figure 20.1). As indicated, petty commodity production and home-based industry avoid various emission laws and codes of safe practice. The haphazard development of repair shops and garages results in toxic waste being dumped into local rivers and the atmosphere. Oil products are particularly dangerous, as heavy rain washes waste into the subsoil and residential water supplies.

The failure of the government to cope adequately with inappropriate housing and other building is reflected in the living environment and in the incidence of environmentally related diseases like gastroenteritis. The principal areas where water-borne infections occur are the western periphery, where population and economic activity are concentrated, and the Gulf of Paria, where wastes are discharged.

A pressing need exists for solutions. Environmental education is required to highlight the effects of human activity in small ecosystems and the value of preserving untouched areas. Conservation areas have been designated in the northern range, part of the central hills, and where coral reefs exist, but bureaucratic problems impair the implementation of such policies. The administrative structure is such that relevant policies and responsibilities are frequently divided between ministries that operate independently. The relevant agencies are often responsible for one resource, a situation which results in fragmented policies and inconsistent goals.

The Informal Versus the Formal Sector and the Green Debate

In the context of national development, the Government of Trinidad and Tobago (1991) has highlighted a number of environmental objectives. Firstly, there is the development of a high-quality living environment, which includes raising health standards through effective resource development that retains the aesthetic value of the landscape. This implies eradication of some informal activity, particularly street trading. Secondly, there needs to be increased economic welfare through the informed use of resources; the high cost of remedial infrastructural works and environmental programmes needs to be channelled into productive expenditure.

The government has its specific targets. Firstly, it wants land to be allocated to its optimum use, which in many cases will be economic. Secondly, it wants to preserve valuable ecosystems like the northern forest and the mangrove swamps. Thirdly, it aims to control the negative effects of development such as pollution, and to encourage environmental awareness among the population. Fourthly, it aims to control informal urban growth by adherence to national planning, with the introduction of new controls where necessary. Particular concern exists about the illegal development of land and the denudation of vegetated and marine habitats. Establishing a green belt around Port of Spain is being considered.

Despite the problems of the informal sector, it has positive features. It is small-scale, local and more indigenous than the large-scale, formal economy that is subject to external control. The local worker has greater links with the environment and should thus take better care of resources. Many of the environmental problems of the informal sector are relatively easy and cheap to address, such as improved rubbish disposal. There are also many activities that involve recycling, and there is minimal waste of resources from advertising and packaging.

The informal sector needs an improved water and sewerage infrastructure and local control of rubbish pollution. This is difficult to achieve with many activities as it implies their registration, thereby making them liable to tax, laws and other regulations. Yet some infrastructural improvements could readily be made and more municipal dumping grounds could be provided for rubbish.

Advocating the growth of the informal sector encourages uncontrolled development and urban degradation. The government of Trinidad and Tobago currently faces this dilemma. Solutions lie in educating small-business entrepreneurs and operators about the environment. Such people should be more concerned than outsiders about the future of the country. Also the government is making training courses and loans available to individuals; environmental awareness should be part of such courses, and loans should depend on following environmental guidelines. In theory, the informal sector can be an eco-friendly form of development.

Conclusion

This chapter highlights environmental problems associated with urban and industrial development in the Caribbean. However, as shown at the United Nations Conference on Environment and Development at Rio de Janeiro in 1992, environmental issues are often pushed aside by Third World countries due to short-term concerns with poverty alleviation or debt repayment. In such instances, the balance between environmental safety and economic progress is unequal, with the environment taking second place. The ability of the informal sector to avoid legislation is often beneficial for job creation, but informality creates problems for planners and environmentalists. As physical planning policies often fail, education is crucial to improving environmental quality, and increased environmental awareness is a first step to solving current problems. The last thirty years have probably seen more environmental degradation in the Caribbean than over previous time, and a pressing need exists for environmentally sustainable local economic development. Yet unless the negative effects of informal development are acknowledged, urban environments will continue to degrade.

References

Beckford, G.L. (1972) *Persistent Poverty: Underdevelopment in Plantation Economies of the Third World*. New York: Oxford University Press.

Besson, J. (1987) A paradox in Caribbean attitudes to land. In J. Besson and J. Momsen (eds), *Land and Development in the Caribbean*. London: Macmillan, pp. 13–45.

Central Statistical Office (1990) *Annual Statistical Digest, Republic of Trinidad and Tobago, Port of Spain*, 37.

Conway, D. (1989) Trinidad and Tobago. In R.B. Potter (ed.), *Urbanisation, Planning and Development in the Caribbean*. London: Mansell, pp. 49–76.

Deere, C., Antrobus, P., Bolles, L., Melendez, E., Phillips, P., Rivera, M. and Safa, H. (eds) (1990) *In the Shadows of the Sun: Caribbean Development Alternatives and US Policy*. Colorado: Westview Press.

Demas, W.G. (1965) *The Economies of Development in Small Countries with Special Reference to the Caribbean*. Montreal: McGill University Press.

Dirks, R. (1978) Resource fluctuations and competitive transformations in West Indian slave societies. In C.D. Laughlan and I.A. Brady (eds), *Extinction and Survival in Human Populations*. New York: Columbia University Press.

Durant-Gonzalez, V. (1985) Higglering: rural women in the internal market system in Jamaica. In P.I. Gomes (ed.), *Rural Development in the Caribbean*. London: Hurst and Company, 103–22.

Farrell, T.M.A. (1978) The unemployment crisis in Trinidad and Tobago: its current dimensions and some projections to 1985. *Social and Economic Studies*, **27** (2), 171–85.

Government of Trinidad and Tobago (1982) *National Physical Development Plan*. Port of Spain: Development Planning Series, Republic of Trinidad and Tobago.

Government of Trinidad and Tobago (1991) *Medium Term Macro Planning Framework 1989–1995*. Port of Spain: National Planning Commission, Republic of Trinidad and Tobago.

Harrison, F. (1991) Women in Jamaica's urban informal economy: insights from a Kingston slum. In C. Mohanty, A. Russo and L. Torres (eds), *Third World Women and the Politics of Feminism*. Bloomington: Indiana University Press, pp. 173–96.

Katzin, M.F. (1960) The business of higglering in Jamaica. *Social and Economic Studies*, **9**, 297–331.

Kirton, C. and Witter, M. (1993) Aspects of the informal economy in Jamaica. In S. Lalta and M. Frekelton (eds), *Caribbean Economic Development: The First Generation*. Kingston: Ian Randle, pp. 280–90.

Le Franc, E. (1989) Petty trading and labour mobility: higglers in the Kingston metropolitan area. In K. Hart (ed.), *Women and the Sexual Division of Labour in the Caribbean*. Kingston: Consortium Graduate School of Social Sciences, University of the West Indies, pp. 99–132.

Lewis, W.A. (1955) *The Theory of Economic Growth*. London: George Allen and Unwin.

Lloyd Evans, S. (1994) *Ethnicity and Gender in the Informal Sector in Trinidad: With Particular Reference to Petty-commodity Trading*. Unpublished PhD thesis, University of London, London.

Lloyd Evans, S. and Potter, R.B. (1992) The informal sector of the economy of the Commonwealth Caribbean. *Bulletin of Eastern Caribbean Affairs*, **17** (3), 26–40.

Lowenthal, D. (1987) Foreword. In J. Besson and J. Momsen (eds), *Land and Development in the Caribbean*. London: Macmillan.

McAffe, K. (1991) *Storm Signals: Structural Adjustment and Development Alternatives in the Caribbean*. London: Zed Books.

Mintz, S.W. (1985) *Sweetness and Power: The Place of Sugar in Modern History*. New York: Viking.

Pearson, R. (1993) Gender and new technology in the Caribbean: new work for women? In J.H. Momsen (ed.), *Women and Change in the Caribbean: A Pan-Caribbean Perspective*. London: James Currey, pp. 287–95.

Phillips, D. (1985) *Women Traders in Trinidad and Tobago*. Port of Spain: Economic Commission for Latin America.

Portes, A., Castells, M. and Benton, L.A. (eds) (1989) *The Informal Economy: Studies in Advanced and Less Developed Countries*. Baltimore: John Hopkins University Press.

Potter, R.B. (1989) *Urbanization, Planning and Development in the Caribbean*. London: Mansell.

Potter, R.B. (1992) *Urbanisation in the Third World*. Oxford: Oxford University Press.

Potter, R.B. (1993) Urbanization in the Caribbean and trends of global convergence–divergence. *Geographical Journal*, **159**, 1–21.

Potter, R.B. (1994) *Low-Income Housing and State Policy in the Eastern Caribbean*. Mona: The Press University of the West Indies, Barbados, Jamaica, Trinidad and Tobago.

Potter, R.B. (1995) Urbanization and development in the Caribbean. *Geography*, **80**, 334–41.

Rampersad, M. (1991) The measurement of the informal sector in Trinidad and Tobago. *Central Statistical Office Research Papers*, Port of Spain.

Richardson, B.C. (1992) *The Caribbean in the Wider World 1492–1992*. Cambridge: Cambridge University Press.

Rojas, E. (1989) Human settlements of the Eastern Caribbean: development problems and policy options. *Cities*, **6**, 243–58.

Safa, H. (1986) Economic autonomy and sexual equality in Caribbean society. *Social and Economic Studies*, **35** (3), 1–22.

Safa, H. (1990) Women and industrialisation in the Caribbean. In S. Stitcher and J. Parpart (eds), *Women, Employment and the Family in the International Division of Labour*. London: Macmillan, pp. 72–97.

Scotland, L.H. (1983) The petroleum windfall, government activity and economic performance in Trinidad and Tobago. Paper presented at First Annual Conference on Trinidad and Tobago Economy, Port of Spain.

Senior, O. (1991) *Working Miracles: Women's Lives in the English-speaking Caribbean*. London: James Currey.

Sheridan, R. (1973) *Sugar and Slavery: An Economic History of the British West Indies 1623–1775*. Baltimore: John Hopkins University Press.

de Soto, H. (1989) *The Other Path*. London: Tauris.

Thomas, C.Y. (1988) *The Poor and the Powerless: Economic Policy and Change in the Caribbean*. London: Latin American Bureau.

Tokman, V. (1989) Policies for a heterogeneous informal sector in Latin America. *World Development*, **17**, 1067–76.

Watts, D. (1987) *The West Indies: Patterns of Development, Culture and Environmental Change since 1492*. Cambridge: Cambridge University Press.

Wolf, E.R. (1982) *Europe and the People without History*. Berkeley: University of California Press.

Worrell, D. (1987) *Small Island Economies: Structure and Performance in the English-speaking Caribbean since 1970*. New York: Praeger.

21

Potential for Conflict? The Proposed Namosi Copper Mine, Viti Levu, Fiji

Frank McShane

Development for Pacific Island countries often involves alliances between overseas private industry and island governments as a basis for exploiting natural resources to fuel economic growth. Such alliances are seen as essential in large-scale mining developments, which require financial and technological inputs far beyond the resources of most Pacific Island countries. The scale and speed of change demanded of the environments, the economies, and the societies of Pacific Islands by the introduction of large-scale mining are always spectacular and often overwhelming. There is a continuing and heated debate concerning the benefits of large-scale mining on Pacific Islands as against its environmental and social costs (Emberson-Bain, 1992). Many of the conflicts associated with mineral development in Melanesia are linked to environmental issues (Grynberg and Nouairi, 1988), particularly loss of resources and degradation of the land. Significant social disruption can also occur.

Fiji is at the threshold of a new era with a proposal to develop a mine in the remote Namosi Province on the main island of Viti Levu. At the time of writing, the project is at the pre-feasibility stage. In this study, an assessment of the project is made based on a Draft Environmental Inception Report prepared by Natural Systems Research, the environmental consultants to the mining company (NSR, 1992), and in the light of the wider debate surrounding open-pit mining on Pacific Islands.

Namosi Province

Namosi Province is located on the southeast side of the volcanic island of Viti Levu in Fiji (Figure 21.1). The area rises to over 1000 m in the Korobasabasaga

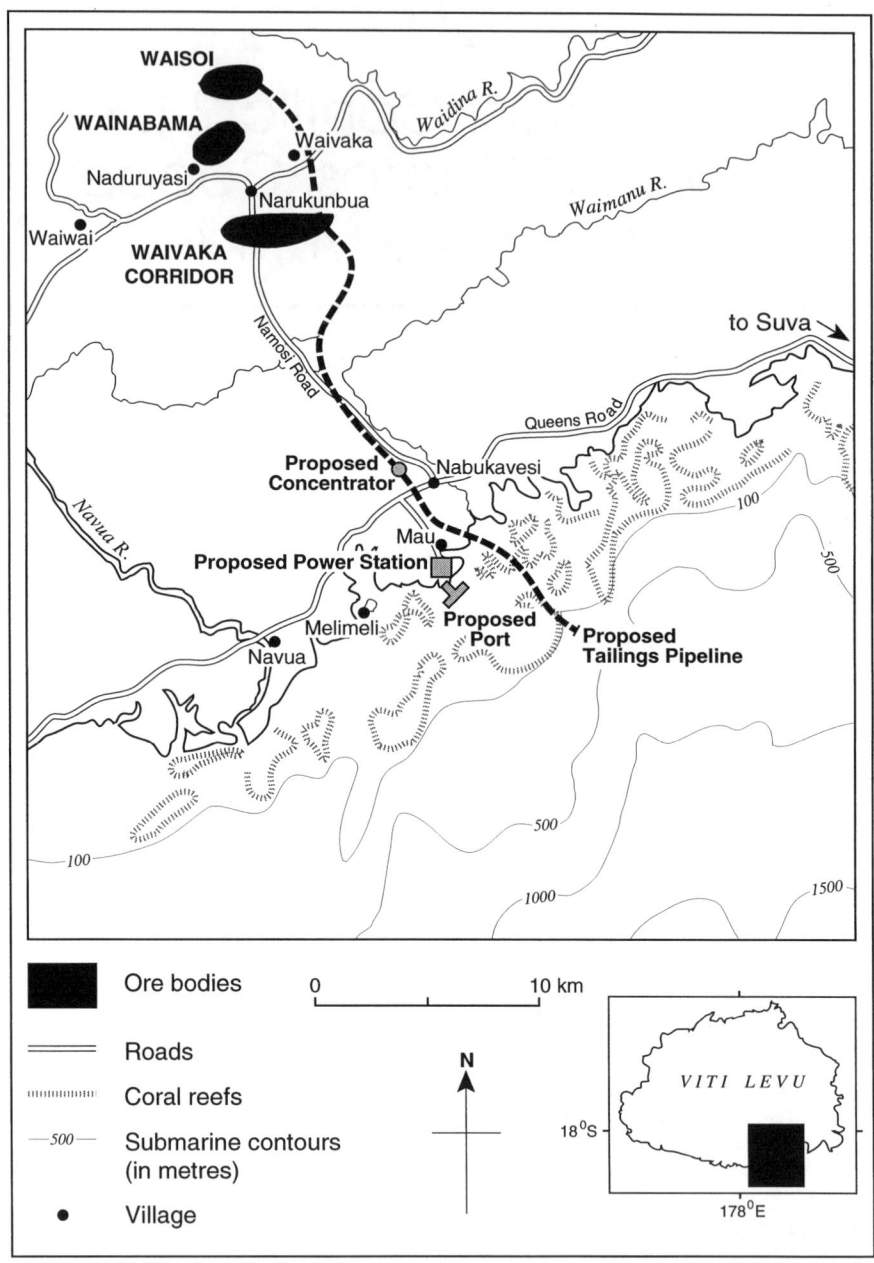

Figure 21.1 Main infrastructure for the proposed Namosi copper mine project, Fiji.

Range. The tropical maritime climate and southeast trade winds bring an average annual rainfall in excess of 4000 mm to the area, most of which falls between December and April. Much of the interior is classified as steepland (with slopes greater than 18°) and is particularly prone to landslides. It is also susceptible to other natural hazards including earthquakes and tropical cyclones. The interior of Namosi is a steep, densely wooded area of dakua (*Agathis vitiensis*) forest, the true botanical diversity of which has not been fully recorded (Watling and Chape, 1992). South of the Namosi area, forest gives way to intermediate montane vegetation, coastal mangroves and marshes, lagoons and offshore reefs. The coastal strip is characterized by tourist developments, which coexist with local agriculture, whose production is consumed locally or sold at markets in Suva and Navua.

The interior of Namosi is serviced by an unsealed road, opened in 1978, that joins the main Suva to Nadi highway on the south coast at Nabukavesi, 24 km from Suva. The proposed mine site is located in the Waisoi area at the base of the Korobasabasaga range. There are several villages nearby, namely, Namosi, Narukunbua, Naduruyasi, Waivaka and Waiwai. Namosi is the largest, with a population of over 500 people. Namosi Province has a population in excess of 5000, most of whom live along the coast. In spite of its proximity to the capital of Suva, Namosi is one of the least developed provinces in Fiji (Rizer *et al.*, 1982), and much of the interior area away from the road remains inaccessible.

History of Exploration at Namosi

A succession of mining companies has investigated the possibility of mining copper from the Waisoi valley area since 1912. Major exploration between 1978 and 1979 by Viti Copper defined a large deposit of 480 million t of copper at a grade of 0.48 per cent, but this was uneconomic to exploit at prevailing prices (Richmond, 1980). Subsequently, in early 1992, a Special Prospecting Licence was granted to Placer Pacific and Placer Dome Limited to prospect the Namosi copper/gold porphyry. These companies are part of the giant Canadian mining conglomerate Placer Dome Inc., currently the fourth largest mining company in the world outside South Africa (Durr, 1990). A further application was made by Placer Exploration Limited in April 1993 to prospect an area of 60 000 ha.

Whilst no public commitment to mine copper at Namosi has been published, there are a number of reasons to believe that Placer Pacific wishes to proceed with the project. Firstly, the company is keen to increase its 'exposure to base metals' and to reduce its overall dependence on gold (Anon., 1989a, 1992). Secondly, Namosi is Placer Pacific's largest prospect and the copper can be extracted comparatively cheaply. Thirdly, the company is spending aggressively, with a commitment of F$5 million (*c*.£2.5 million at 1994 rates) for the pre-feasibility survey. Fourthly, the investment climate in Fiji is currently considered to be more favourable than elsewhere in the South Pacific.

Infrastructural Requirements of the Project

Everything about the Namosi project is on a gigantic scale and has significant implications for the local environment and people. There will be two large pits dug in the Waisoi valley, one 2000 m in diameter, the other 1200 m. Both will be 200 m deep. The project will require the diversion of the Waisoi creek to prevent flooding of the pits, and the establishment of a waste rock dump in the Waisoi valley (Placer Pacific, 1993).

The project will make huge demands in terms of manpower and infra-structure, including an 85 to 135 megawatt power station, a demand in excess of that currently supplied to the national grid at times of peak consumption. The power station will be situated on the south coast between Suva and Navua. The project will also require upgrading of the road to Namosi and the construction of a new port for export of the mine products, imports of fuel for the power station and of other items. A camp for some 3000 workers will be required during the mine construction phase. There will also need to be a concentrator at the coast, a slurry pipeline from the crushing mill to the concentrator, and a long sea outfall to carry waste tailings offshore to the submarine slope. The operation will draw large quantities of water from the ocean near the Beqa passage for cooling at the power station. Water for the crushing and grinding processes at the mine and the concentrator will probably be pumped from the Navua river.

Namosi: an Environment of Concern?

The Namosi project involves mining very low-grade ore in large quantities. To be viable, the project will require a financial investment in infrastructure and services on a scale unparalleled in Fiji. While the mine will consume a non-renewable resource, it will add to the capital base of the country through royalties, direct and indirect taxes, foreign-exchange earnings, service industries, local business developments and employment. In terms of local benefits, there is the probability of 1000 jobs, provision of roads, schools, health services and housing, and also gains from compensation and equity (Keith-Reid, 1993).

Yet large-scale mining has been a contentious issue in Melanesia. The idle mine at Bougainville stands like an icon of another time: a decade of naiveté for Papua New Guinea. The tale of misplaced trust and mismanagement has been recounted by Grynberg and Nouairi (1988), but Bougainville is not alone. Other mine operators in Papua New Guinea strive constantly to convince their detractors that, while the mines are not sustainable, they can be manageable. Yet the mines at Misima and Porgera, and those proposed at Mount Kare and Lihir, have attracted attention and controversy since their inception. Emberson-Bain (1992) has summarized the undesirable consequences of mining activities in the South Pacific, including pollution and loss of land, the creation of local dependency cultures, shifts to imported foods, tensions associated with immigrant labour, health problems, the breakdown of the extended family, and the erosion of women's status in the community. In this context, many aspects of the Namosi project require critical consideration. The particular emphasis here is on the management of waste products and the capacity of the local area to accommodate the project.

Waste Management at Namosi

Two important waste management issues arise in open-cast mining projects like Namosi, namely, the generation of waste rock and the disposal of tailings.

Generation of Waste Rock

During the mining process not all the rock extracted will contain usable copper. At Namosi, approximately 100 000 t of waste rock will be produced each day for every 100 000 t of ore mined. If the waste rock contains residual ferrous iron or sulphides, there is then the potential for oxidation of these compounds, e.g. conversion of sulphides to acidic sulphates, which can form acid solutions that affect both surface and ground waters (Hore-Lacy, 1992). Acid solutions of this kind can dissolve heavy metals and become a serious problem, particularly under moist and warm conditions. During mining, a much greater area of potentially acid-forming rock is exposed than is naturally the case.

Mining companies possess the technology to deal effectively with acid drainage problems by capping acid-waste dumps with an impermeable layer to prevent water and oxygen reaching the waste rock. In the environmental consultants' report at Namosi (NSR, 1992), specific amelioration measures are not addressed. The report is rather vague on the subject, referring simply to 'progressive rehabilitation'.

Waste Disposal

Of each tonne of rock sent to the crushing mill at Namosi, only 0.45 per cent will be copper. From the ore mined each day, approximately 500 t of copper will be shipped as a 25 per cent concentrate in 2000 t batches, leaving 98 000 t of tailings, composed of sand, silt and clay particles. According to the environmental consultants, the most efficient way of disposing of this residue is in the deep ocean using a system called submarine tailings disposal. This option was selected because it involves less land alienation, reduces the danger of toxic hazards on land, and allows the ocean salinity to neutralize the potential acidity of the waste. It is assumed that the biologically productive upper water column will not be degraded, but this is unproven. The most compelling point made by the environmental consultants in favour of submarine tailings disposal at Namosi is that it removes the need to build a tailings dam in a geologically unstable region, thus eliminating the danger of catastrophic dam failure. Advocates of tailings dams have argued that, if the ground were cleared to bedrock, a common practice in strip mining operations, then stable foundations could be laid for a tailings dam even in such a region, but costs would be much higher than for submarine tailings disposal.

The environmental consultants cite the operations at Island Copper mine on Vancouver Island in Canada and at Misima mine in Papua New Guinea as examples of good submarine tailings disposal practice. The tailings at Island Copper are discharged into the calm waters of a fjord and few impacts have been recorded on local commercial marine species. Ellis (1989, p. 93) indicates that 'in general, coastal systems recover quickly from the smothering impacts of detoxified mine wastes dumped at sea once mining has ceased ... a few years for abundant opportunist species, or several decades for climax organisms'. However, the situation is entirely different off the coast of Namosi Province, where the characteristics of the water body, the incidence of cyclonic events and earthquakes, the proximity of fragile reefs and tourist attractions, and the use of the area by local fishermen require special consideration. Contingency planning for the impacts of a potential rupture of the tailings pipeline is also required.

The Misima gold mine in the Milne Bay Province of Papua New Guinea produces 16 500 t of tailings per day. Submarine tailings disposal has been largely successful, and video film shows the tailings moving into deep water as a coherent fluid stream (Placer Pacific, 1993). Yet some environmental costs have been associated with the mine and, according to Hughes (1989, p. 5), 'soft wastes pollute the sea for 9 km along the coast'. The government of Papua New Guinea has also been highly critical of the operation, stating that pollution from the mine has caused a shortage of drinking and cooking water in nearby creeks and posed a serious health threat (Anon., 1989b).

Ellis (1989) states that the major impacts of submarine tailings disposal are sea-bed smothering and contamination, causing loss of benthic fauna. The implications of this for Namosi Province depend on the role that these fauna play in the food chain of near-surface organisms. It is clear that 'there must be little or no risk at the disposal site of ... tailings upwelling back into the shallow water ... [but] there is unfortunately virtually no environmental data in the public domain about the existing cases' (Ellis and Ellis, 1993, p. 4). If the tailings disposed of at sea degrade the coral reefs near Pacific Harbour, which is a prime recreation spot for locals and tourists, the public outcry will seriously undermine confidence in the Namosi project. Likewise, any impact on the inshore subsistence fisheries, upon which the coastal population rely, will create adverse publicity.

Other Environmental Impacts at Namosi

Other potential impacts exist in the Namosi area. Many are the unavoidable accompaniments of the mining process, but others are preventable. The loss of native forest, soil erosion and increased sedimentation fall into the former category. Contamination of local water sources, pollution from the power station, including thermal pollution, contamination of the coastal area as a result of leakages during discharge and loading operations, and the environmental and social impacts of the construction activities are in the latter category. The specific issues considered here are chemical contamination and increased sediment discharge.

Chemical Processes

After the waste has been discarded, the ore is sent to the mill for crushing, after which the copper is separated out by flotation. Flotation involves blowing air through a slurry containing a reagent that causes the copper to collect as a froth on the top of a tank. Thereafter, gravity separation provides a concentrate that contains 20 to 30 per cent copper. The concentrate will be shipped abroad for smelting and further refinement, prior to being turned into copper castings.

It seems likely that any pollution at the mine site and processing works will enter the surface water and ground water systems. The potential for pollution at the processing works will be of great concern to the people of Lobau, Mau, Melimeli, Veivatuloa and Naqara in respect of both fisheries and the coast itself. However, the decision to use only crushing and flotation methods to separate off the copper and any gold, and not chemical leaching involving cyanide, circumvents the major potential impacts of chemical spillage associated with mining operations. Further, large-scale chemical pollution is not necessarily an accompaniment of mining, although the literature abounds with examples (Brodie, 1990). Island Copper in Canada has been successful in avoiding detrimental pollution through comprehensive control and monitoring procedures, but other mines, like Marcopper at Mardinduque island in the Philippines, have had disastrous and far-reaching pollution impacts (Gourlay, 1992).

Ellis (1989) states that the pollution monitoring and impact assessment conducted at Island Copper were one of the most extensive ever conducted and that the mine has continually met pollution control standards. On one occasion when fish mortalities occurred, the cause was traced to a new frothing agent used in the milling process which was quickly withdrawn. Should the Namosi project go ahead, effective pollution control procedures appropriate to a tropical coastal environment must be a priority for the government and the mining company.

Increased Sediment Discharge

The Namosi mine will extract 36 million t of rock annually which will generate a significant additional sediment load. Namosi and the adjacent villages use the surrounding freshwater habitat, particularly the upper Waidina river and its tributaries, as a source of fish, prawns and eels. A census conducted by the Namosi Joint Venture (1979) revealed that, from 78 respondents, 80 per cent fished at least once a week and 51 per cent fished five to six times a week. People in downstream areas conducted more fishing trips than those upstream and the 'amount of fish/prawns/eels caught was unexpectedly high' (Namosi Joint Venture, 1979, p. 60). At present, the effects of increased sediment load on the fisheries resource is unknown. However, several groups surveyed by the Namosi Joint Venture attributed a recent decrease in fish abundance to a heavy build-up of sediment in the river as a result of bulldozing for the road and of exploration drilling. If mining activities expand, then increased sediment load is likely further to degrade the fisheries resource.

Any increase in sediment production from Namosi will also exacerbate flooding in downstream areas during tropical cyclones and other storms by filling river channels and reducing their discharge capacity. The environmental consultants, in co-operation with the Hydrological Unit of the Public Works Department, are currently monitoring sediment discharge in the Waidina and adjacent river catchments. As anticipated, natural discharge rates and loads are extremely high, particularly during cyclones. Levels of sediment discharge under the Nausori bridge reached an estimated 25 million t over four days in January 1993 during Cyclone Kina (Hargreaves, 1993). Increased sediment loads will clearly intensify the flood hazard.

Environmental Management in Fiji

The sheer size of the Namosi project warrants a critical reassessment of the adequacy of current environmental legislation in Fiji. According to Watling and Chape (1992, p. 128), 'Fiji's environmental laws are many and varied, a relic of the colonial period when environmental problems were limited and clearly sectorial. At least 25 Acts have some important role to play in what is today perceived as environmental management.' Although an Environmental Management Committee has existed since 1980 and a small Environmental Unit was created in 1989, many of Fiji's environmental laws are ineffective through lack of enforcement, technical expertise and funding.

The main law governing the environmental management of mines is the Mining Act, which allows for compensation and restoration of land damaged by mining (Government of Fiji, 1966). The maximum fine of F$200 (*c*.£100) or six months in jail is clearly inadequate. Numerous other acts, referring to water pollution, catchment management, public health, fisheries and native land trust also have implications for the environmental management of large mining projects. The Rivers and Streams Ordinance Act of 1882, for example, is the main instrument for the control of river pollution. It is outdated and prone to misinterpretation. No ground water legislation exists.

Since 1981, there has been a specific requirement to undertake an environmental impact assessment as a condition of any mineral exploration project in Fiji. A government commitment was also given to safeguard the environment and improve pollution monitoring (Government of Fiji, 1981). Placer Pacific has undertaken to conduct an environmental impact assessment. Yet according to Roger Moody (personal communication, 1993) of Minewatch, the environmental consultants' statement that 'at any time the level of activity on environmental investigation is commensurate with Placer's confidence in the project' (NSR, 1992, p. 23) gives some cause for concern as to the company's environmental commitment.

Capacity of the Local Area to Accommodate the Namosi Project

Environmental concerns for the land are writ large in the history of mining in Melanesia. They are intimately associated with issues of land ownership and compensation. Connell and Howitt (1991, p. 4) comment that 'for indigenous people, land is clearly central to social identity'. In Melanesia, culture, tradition and the requirements of subsistence living involve people intimately with the land. The macro-economic advantages of mining so often promoted by governments may mean little to a future generation that shares few of the benefits and is deprived of its land. Filer (1990) observes that in Bougainville the people associate the mining damage done to their environment as symbolic of the damage done to their society.

To ethnic Fijians the importance of land or *vanua* is summarized by Ravuvu (1983, p. 70):

> The Fijian term *vanua* has physical, social and cultural dimensions which are interrelated. It does not mean only the land area one is identified with, and the vegetation, animal life and other objects on it, but it also includes the social and cultural system ... the people, their tradition and customs, beliefs and values.... .It provides a sense of identity and belonging.... To most Fijians, the idea of parting with one's *vanua* or land is tantamount to parting with one's life.

The land issue is complicated further because, in many Melanesian countries that have inherited colonial administrative and legal frameworks, landowners are bound by constitutional rather than traditional regulation of mineral resource rights. In Fiji, the landowners have customary rights to the land, but the Mining Act indicates that the ownership of minerals is vested in the State (Government of Fiji, 1945, 1966). A significant amendment to the Act, in 1990, specified that

> any royalties or proceeds received by the State in respect of any minerals extracted from the land ... become payable to the owners of the surface of that land ... subject to the right of the State to retain such proportion of any such royalties or proceeds as may be approved by the Cabinet from time to time. (Government of Fiji, 1990, p. 1)

Mining affects traditional relationships with the land in various ways, including (a) the direct loss of land for open pits; (b) changes in the use of land for the construction of a work camp, roads and other infrastructure; (c) the indirect loss of land and resources through pollution and other degradation; (d) the appropriation of land within a lease area which may or may not be used, but pre-empts other uses; and (e) the direct expropriation of land.

In the past, Viti Copper negotiated financial compensation based on the activity and use of land during the exploration stage, which involved small-scale environmental disturbance and some imposition on local people. If Placer Pacific proceeds with the current project, compensation will be much greater and will increase with the scale of mining. The relevant negotiations will be carried out by the Native Land Trust Board on behalf of local landowners. Placer Pacific has prepared a Draft Compensation Agreement in which it 'acknowledges that if the Project is developed certain traditional areas of the land, freshwater creeks and rivers, and marine environment will be affected' (Placer Pacific, 1993, p. 1). During the period for which it holds a Special Mining Lease, the company has agreed to pay compensation to the landowners for damage to the surface of the land including crops and trees, disturbance of surface rights of way and riverine water, and related impacts.

Compensation has also been proposed for habitual users of creeks within the project area and for any water discoloration and deleterious effects to aquatic flora and fauna. Constant alternative sources of water will be supplied where necessary. The statement that these will be provided 'as far as is possible . . . before the creek is discolored' (Placer Pacific, 1993, p. 1) suggests that they will be in place of mitigatory action. The company has also made provision for marine compensation, including interference with community access, occupation of the marine areas, and damage to or loss of coastal marine areas, including fishing access and catches.

Yet, according to Hughes and Sullivan (1992, p. 6),

> two major reasons for conflict within landowning communities and between these communities and the mining companies and governments, are the trauma that results from the permanent loss of land because of mining, and disputes about the ways in which revenue and other benefits are distributed.

Further, Connell and Howitt (1991) assert that it is too easy to argue that the 'problem' of environmental degradation and land loss is solved by replacing rights with money. It remains to be seen whether Placer Pacific can identify a compensation scheme that will equitably recompense all members of the community affected by the mine for their loss of resources and land access. Elsewhere, failure to do so has created division and conflict between landowners, mining companies and governments.

Conclusion

Many of the potential environmental problems at Namosi can be resolved by the application of good environmental and resource management techniques, although some impacts cannot be adequately assessed until the mine is operational. For the villagers of Namosi, the important issues are those of environmental degradation and of compensation for loss of land access and resources. Although methods for compensating landowners for environmental degradation and social impact have doubtless become more elaborate, it remains to be seen whether multinational mining companies are any better equipped to empathize with those whom their decisions affect than was the case in Bougainville.

Many people in remote areas, although acutely aware of the progress and, in some cases, the wealth offered by large-scale mining developments, are often uncomprehending of the potential environmental and social changes accompanying such projects. Mining in remote areas brings the difference between the traditional worldview and the ethos of multinational corporate culture and shareholder satisfaction into sharp focus. The result is often confrontation, each side believing that their needs and objectives have been understood by the other, yet only to be disillusioned by later events.

Acknowledgements

Roger Moody of Minewatch, London kindly provided information. Professor Eric Waddell, Dr Paddy Nunn, Dr Bill Aalbersberg, Andrew Crosby and Heather McShane contributed valuable comments and advice.

References

Anon. (1989a) Placer Dome sets targets. *Mining Journal*, 8 December, p. 471.

Anon. (1989b) Placer told to clear up. *Pacific Islands Monthly*, July, p. 39.

Anon. (1992) Want it right? Do it yourself. *Energy and Mining Journal*, December, p. 11.

Brodie, J.E. (1990) *The State of the Marine Environment in the South Pacific*. United Nations Environment Program. Regional Seas Report No. 127, South Pacific Regional Environment Program Topic Review 40.

Connell, J. and Howitt, R. (1991) Mining dispossession and development. In J. Connell and R. Howitt (eds), *Mining and Indigenous Peoples in Australasia*. Sydney: Sydney University Press.

Durr, B. (1990) Placer Dome puts a shine on its activities. *Financial Times*, London, 13 October.

Ellis, D. (1989) *Environments at Risk: Case Histories of Impact Assessment*. Berlin: Springer Verlag.

Ellis, D. and Ellis, K. (1993) Very deep submarine tailings disposal. Unpublished paper.

Emberson-Bain, A. (1992) Sustaining the unsustainable? Assessing the impact of mining in the Pacific, with particular reference to women. Unpublished paper.

Filer, C. (1990) The Bougainville rebellion, the mining industry and the process of social disintegration in Papua New Guinea. *Canberra Anthropology*, **13**, 1–39.

Gourlay, R. (1992) Philippines orders mine closure because of pollution problem. *Financial Times*, London, 22 April.

Government of Fiji (1945) An ordinance relating to prospecting and mining precious metals and other minerals. In *Laws of Fiji*. Suva: Government House, p. 1326, Cap. 127 and later amendments.

Government of Fiji (1966) The Mining Act. In *Laws of Fiji. 1966*. Suva: Government House.

Government of Fiji (1981) *Eighth Development Plan, 1981–1985*. Suva: Government Publishers.

Government of Fiji (1990) *The Constitution of Fiji*. Suva: Government Publishers.

Grynberg, R. and Nouairi, J. (1988) The environmental costs of development in PNG: the case of Bougainville copper. Unpublished report, Department of Economics, University of the South Pacific, Suva.

Hargreaves, I. (1993) Environmental monitoring for the proposed Namosi mine. Talk to Chemical Society of the South Pacific, 17 July, University of the South Pacific, Suva.

Hore-Lacy, I. (1992) *Mining and the Environment*. Adelaide: Australian Mining Industry Council.

Hughes, P.J. (1989) *The Effects of Mining on the Environment of High Islands: A Case Study of Gold Mining on Misima Island, Papua New Guinea*. South Pacific Regional Environment Case Study No. 5., Noumea.

Hughes, P.J. and Sullivan, M. (1992) *The Environmental Effects of Mining and Petroleum Production in Papua New Guinea*. Port Moresby: Natural Resource Series, University of Papua New Guinea.

Keith-Reid, R. (1993) Hope in the hills: Namosi and the minerals that would make it a rags to riches story. *Islands Business Pacific*, May, 36–40.

Namosi Joint Venture (1979) *Environmental Progress Report, No. 2, January 1st 1978 – December 31st 1978*. Suva: Namosi Joint Venture.

NSR (1992) *Namosi Draft Environmental Inception Report*. Victoria: Natural Systems Research and Placer Pacific Exploration Limited.

Placer Pacific (1993) *Current Status of the Namosi Project as of March 31st 1993*. Suva: Placer Pacific Namosi.

Ravuvu, A. (1983) *Vaka i Taukei: The Fijian Way of Life*. Suva: Institute of Pacific Studies, University of the South Pacific.

Richmond, R.N. (1980) Case history of mineral development in Fiji: the Namosi copper project. Paper presented at Workshop on Mineral Policies to Achieve Development Objectives, 9–13 July, East–West Centre, Hawaii.

Rizer, J.P., Lin, J., Waqavonovono, M., Saumatua, S. and Majoram, A.G. (1982) *The Namosi Copper Mine: A Study in Assimilation Planning*. Suva: University of the South Pacific.

Watling, D. and Chape, S.P. (eds) (1992) *Environment Fiji – the National State of the Environment Report*. Gland: International Union for Conservation of Nature.

PART VI

Conclusion

22

Land Degradation in the Tropics: The Way Ahead

Michael J. Eden and John T. Parry

Even the most carefully planned land developments in the tropics, whether agricultural or urban–industrial, involve a degree of environmental degradation. It is clearly desirable to ensure that such degradation, or 'creative destruction' as it is termed by Johnson and Lewis (1995), is no more than commensurate with the benefits of the developments involved. Equally, it is desirable to minimize the degradation that arises from careless or unsustainable land exploitation. In either case, attention has to be paid to direct or 'on-site' degradation and to any indirect degradation of other habitats. The rationale behind this approach is that, over time, it will conserve environmental resources, facilitate sustainable exploitation, and minimize social degradation. Yet sensible as such 'ecodevelopment' may seem, it easily founders in the real tropical world of increasing human populations, over-exploited resources and imperfect governance. The question that arises is how far effective constraints on land degradation can really be exercised, given the developmental and other pressures that currently exist in many parts of the tropics.

Effective management of land degradation in the tropics depends, as elsewhere, on a variety of factors. Of major importance are, firstly, the availability of appropriate technical strategies for land development and for managing associated degradation, and, secondly, a willingness and ability on the part of governments to implement such strategies. Numerous technical issues remain to be resolved, notably in relation to the environmental feedbacks of large-area land cover changes and to processes of land restoration and vegetation renewability, but these issues are less problematic than the task of persuading governments to implement policies for managing land degradation. The challenges can only be met by a combination of appropriate technical strategies and clear-sighted action on the part of national governments.

It is important that substantial attention is paid to specific degradations, as has been done in most chapters in this volume. However, a further issue arises with the adoption of the broader and more integrative concept of land degradation, namely, the *relative* seriousness of specific degradations. This needs to be clarified prior to formulating national or regional environmental policies that are modulated by the finite, and often limited, financial resources available to governments in the tropics. In this context, specific degradations need to be considered in terms of 'net degradation', with due attention being paid to restoration as well as degradation processes, whether human or natural (Blaikie and Brookfield, 1987).

Some degradation issues, not least desertification and deforestation, have acquired high profiles in recent decades. This is due to the apparent seriousness of the environmental and human impacts involved, but it also reflects the considerable attention paid to the issues by scientists, non-governmental organizations and the media. However, in terms of net degradation, it is arguable that the impacts of desertification and deforestation have been exaggerated and are less threatening than has at times been claimed; for example, the resilience of the Sahel zone (Helldén, 1991; Mace, 1991) and the renewability of Amazonian rain forest (Eden, 1994) have commonly been underestimated. More recent concern over loss of biodiversity, which acquired political credibility at the United Nations Conference on Environment and Development at Rio de Janeiro in 1992, shows a useful sharpening of the scientific perspective on land degradation, but the issue itself is less tangible and newsworthy, and, for all its technical validity, may not so easily develop the high profile that encourages governments to act. Equally, the emerging concern over the degradation of wetlands, which are reported to 'rival tropical rain forests for priority on the world conservation agenda' (Maltby and Dugan, 1994, p. 29), is based on only limited investigation in the tropics. The inherent value of wetlands is clearly high, but, as yet, is less widely acknowledged.

What then are the real priorities in land degradation, and where should policy-making and investment focus? The various degradations mentioned above are very serious, but many would argue that soil erosion, with its direct impact on land productivity and its effective irreversibility on a human timescale, constitutes a more immediate and critical threat to economic activity and human welfare. Equally, urban–industrial degradation which affects the health and welfare of a substantial and increasing proportion of the tropical population arguably warrants much more attention than it has hitherto received. The relative seriousness of these broad-based degradations has not been adequately evaluated, nor indeed have sufficient data been assembled to allow such an evaluation. The task awaits attention, and its outcome is relevant to future environmental policy-making.

Even if sound environmental priorities and policies are established for land degradation, persuading governments to implement the policies is a mammoth task. Governments in the tropics, as elsewhere, have their own agendas, and, although many governments formally acknowledge the importance of environmental issues, they are often unwilling or unable to accord them the priority or sensitive treatment they require. In such circumstances, there has been a recurrent tendency to emphasize the value of a grassroots approach to land development and to stress the role of local people, local knowledge and local initiatives. Examples of this viewpoint include Richards's (1985) espousal of a peasant-focused 'indigenous agricultural revolution' for West Africa, and Hecht and Cockburn's (1989) promotion of a 'socialist ecology' as the only strategy that can save Amazonia and its inhabitants. The arguments deployed are cogent enough, yet, in the face of such pervasive degradations as tropical deforestation, soil erosion or urban–industrial pollution, it is self-evident that national governments cannot be bypassed and that, irrespective of any grassroots input, governments themselves have ultimately to be persuaded of the need for or be pushed into a real commitment to degradation management.

Effective management of degradation depends on many factors, of which the most fundamental is a degree of political stability and competent governance in the country concerned. The optimal technical strategy for controlling degradation in a particular place is always open to debate, but it is evident in the tropics, as elsewhere, that control can be precluded by the instability or incompetence of government. It is not a question of strategies or policies, but of governance itself. Although most countries are sovereign, any significant improvement in governance, if it occurs, is commonly a response to external political or economic pressures as well as internal forces. Equally, in countries where governments are in full control, but, for whatever reason, fail to devote adequate resources to sustaining development and managing degradation, external pressures are again often applied.

For better or worse, the international agencies and developed world governments that have habitually been involved in aiding tropical land development thus continue to exert a major influence. Both the nature of and the motives behind their efforts have been criticized over the years (Hayter, 1971; Redclift, 1984; Hayter and Watson, 1985; Adams, 1990), with current concern typically being expressed about the side-effects of economic adjustment programmes and the low priority given to social and environmental concerns (Rich, 1994). Even so, continuing tropical land development demands significant capital investment, implying a persisting, if at times stressed, relationship between external funding sources and tropical governments. In spite of significant private capital flows to 'emerging markets' in the tropics, external public funds remain crucial to tropical land development, particularly in poorer countries where land degradation is most severe.

The problem is that such investment, public or private, may promote land development but carries with it major environmental and social costs that are frequently ignored or underestimated. In effect, the market fails to acknowledge the social value of the environment (World Bank, 1992). This complex issue is beginning to be explored in wetland and forest habitats (Barbier, 1993; Adger *et al.*, 1995), but investment urgently needs to be more appropriate in environmental and social terms.

Non-governmental organizations and the media also have a significant role to play in influencing governmental policies and performance. Much external publicity has been generated over some tropical degradations and has served to increase the environmental accountability of many tropical governments. External non-governmental organizations have also targeted the policies of international funding agencies and developed world governments, thereby indirectly influencing the course of tropical land development (Sasson, 1990; Rich, 1994). As yet non-governmental organizations and the media within tropical countries have been less influential, but their importance is gradually increasing, and, in the future, will focus attention on government accountability and contribute to environmental policy-making (Smiet, 1993; Silva, 1994). Progress in all these directions is patchy and slow, since unsustainable land development and inequitable wealth distribution, with their adverse implications for the environment, often gain support from powerful political coalitions in tropical countries (Silva, 1994). Yet even the most reactionary governments eventually yield to external pressures or internal forces, and, particularly as specific degradations incur increased socio-economic costs, political responses are inevitable. This is a process in which non-governmental organizations and the media can be influential.

Many broader factors also impinge on the long-term prospects for land degradation. Critical among these is human population growth, which has lately received increasing global attention, as shown at the International Conference on Population and Development in Cairo in 1994. While the linkage between population growth and tropical land degradation is by no means straightforward, existing rates of growth frequently compromise efforts that are made to control degradation (Myers, 1992). That population growth is now widely recognized as an independent and manageable variable, rather than one whose control depends on economic growth, can only aid long-term efforts to limit tropical land degradation. On that basis and assuming due attention is paid to social as well as environmental conditions, a strategy for confronting the hazard of land degradation begins to emerge.

References

Adams, W.M. (1990) *Green Development. Environment and Sustainability in the Third World.* London: Routledge.

Adger, W.N., Brown, K., Cervigni, R. and Moran, D. (1995) Total economic value of forests in Mexico. *Ambio*, **24**, 286–96.

Barbier, E.B. (1993) Sustainable use of wetlands. Valuing tropical wetland benefits: economic methodologies and applications. *Geographical Journal*, **159**, 22–32.

Blaikie, P. and Brookfield, H. (1987) *Land Degradation and Society*. London: Methuen.

Eden, M.J. (1994) Environment, politics and Amazonian deforestation. *Land Use Policy*, **11**, 55–66.

Hayter, T. (1971) *Aid as Imperialism*. Harmondsworth: Penguin Books.

Hayter, T. and Watson, C. (1985) *Aid. Rhetoric and Reality*. London: Pluto Press.

Hecht, S. and Cockburn, A. (1989) *The Fate of the Forest. Developers, Destroyers and Defenders of the Amazon*. London: Verso.

Helldén, U. (1991) Desertification – time for an assessment? Ambio, **20**, 372–83.

Johnson, D.L. and Lewis, L.A. (1995) *Land Degradation: Creation and Destruction*. Cambridge: Blackwell.

Mace, R. (1991) Overgrazing overstated. *Nature*, **349**, 280–1.

Maltby, E. and Dugan, P.J. (1994) Wetland ecosystem protection, management, and restoration: an international perspective. In S.M. Davis and J.C. Ogden (eds), *Everglades. The Ecosystem and its Restoration*. Delray Beach: St Lucie Press, pp. 29–46.

Myers, N. (1992) Population/environment linkages: discontinuities ahead. *Ambio*, **21**, 116–18.

Redclift, M. (1984) *Development and the Environmental Crisis. Red or Green Alternatives*. London: Methuen.

Rich, B. (1994) *Mortgaging the Earth. The World Bank. Environmental Impoverishment and the Crisis of Development*. London: Earthscan Publications.

Richards, P. (1985) *Indigenous Agricultural Revolution*. London: Hutchinson.

Sasson, A. (1990) *Feeding Tomorrow's World*. Paris: United Nations Educational, Scientific and Cultural Organization.

Silva, E. (1994) Thinking politically about sustainable development in the tropical forests of Latin America. *Development and Change*, **25**, 697–721.

Smiet, A.C. (1993) Tropical forestry in the 21st century: limitations and opportunities. *Ambio*, **22**, 50–1.

World Bank (1992) *World Development Report 1992. Development and the Environment*. New York: Oxford University Press.

Index